Lecture Notes in Electrical Engineering

Volume 573

The book series *Lecture Notes in Electrical Engineering* (LNEE) publishes the latest developments in Electrical Engineering - quickly, informally and in high quality. While original research reported in proceedings and monographs has traditionally formed the core of LNEE, we also encourage authors to submit books devoted to supporting student education and professional training in the various fields and applications areas of electrical engineering. The series cover classical and emerging topics concerning:

- Communication Engineering, Information Theory and Networks
- Electronics Engineering and Microelectronics
- Signal, Image and Speech Processing
- Wireless and Mobile Communication
- Circuits and Systems
- Energy Systems, Power Electronics and Electrical Machines
- Electro-optical Engineering
- Instrumentation Engineering
- Avionics Engineering
- Control Systems
- Internet-of-Things and Cybersecurity
- Biomedical Devices, MEMS and NEMS

For general information about this book series, comments or suggestions, please contact leontina. dicecco@springer.com.

To submit a proposal or request further information, please contact the Publishing Editor in your country:

China

Jasmine Dou, Associate Editor (jasmine.dou@springer.com)

India

Swati Meherishi, Executive Editor (swati.meherishi@springer.com)
Aninda Bose, Senior Editor (aninda.bose@springer.com)

Japan

Takeyuki Yonezawa, Editorial Director (takeyuki.yonezawa@springer.com)

South Korea

Smith (Ahram) Chae, Editor (smith.chae@springer.com)

Southeast Asia

Ramesh Nath Premnath, Editor (ramesh.premnath@springer.com)

USA, Canada:

Michael Luby, Senior Editor (michael.luby@springer.com)

All other Countries:

Leontina Di Cecco, Senior Editor (leontina.dicecco@springer.com)
Christoph Baumann, Executive Editor (christoph.baumann@springer.com)

**** Indexing: The books of this series are submitted to ISI Proceedings, EI-Compendex, SCOPUS, MetaPress, Web of Science and Springerlink ****

More information about this series at http://www.springer.com/series/7818

Paolo Di Barba · Slawomir Wiak

MEMS: Field Models and Optimal Design

Springer

Paolo Di Barba
Department of Electrical, Computer
and Biomedical Engineering
University of Pavia
Pavia, Italy

Slawomir Wiak
Institute of Mechatronis
and Information Systems
Lodz, Poland

ISSN 1876-1100 ISSN 1876-1119 (electronic)
Lecture Notes in Electrical Engineering
ISBN 978-3-030-21495-1 ISBN 978-3-030-21496-8 (eBook)
https://doi.org/10.1007/978-3-030-21496-8

This Springer imprint is published by the registered company Springer Nature Switzerland AG
The registered company address is: Gewerbestrasse 11, 6330 Cham, Switzerland

Contents

Chapter 1
Introduction

This introductory monograph presents a broad overview of methods of both analysis and synthesis of Micro Electro Mechanical Systems (MEMS) and devices, mainly addressed to graduate students and young researchers in the area of electrical and computer engineering as well as mechatronics. Throughout the book each theoretical concept is illustrated by means of case studies, following a problem-solving approach and never forgetting that the engineering task is just that of formulating and solving problems in a computational fashion. Having this in mind, the authors have collected the experience they have accumulated while teaching electromagnetics and electromechanics at various levels and in different countries, in this book, which is intended to be valid for an international audience.

In particular, the MEMS devices the various case studies are focused on, are analyzed by means of distributed models (i.e. field models) which allow to represent the physical layer i.e. the internal reality which takes place within materials forming the device. In contrast, the traditional approach to analysis is based on lumped-parameter models (i.e. circuit models), which rely on the assumption of disregarding space effects. This twofold approach and its implications at the computing level is described in Chap. 2.

Following the field-model approach, in Chaps. 3, 4 and 5 an overview of electrostatics, magnetostatics and steady-state conduction is presented, respectively. Methods to simulate the mechanical effects which take place in the field, i.e. forces and torques of electromagnetic origin acting on a structure, are accordingly illustrated. In turn, an introduction to multiphysics problems is developed in Chap. 9. On the other hand, the lumped parameter approach is also exploited in the book: the integral parameters characterizing a MEMS device (e.g. the equivalent capacitance), can be reliably computed just starting from field analyses; accordingly, in Chap. 6, the field-circuit approach is proven and discussed.

In order to solve field models in a numerical fashion, many methods can be used; however, the most popular one nowadays is the Finite Element Method (FEM) thanks to its reliability and effectiveness: because most of the case studies presented in the book are modelled by means of FEM, basics of FEM are summarized in Chap. 8.

P. Di Barba and S. Wiak, *MEMS: Field Models and Optimal Design*, Lecture Notes in Electrical Engineering 573, https://doi.org/10.1007/978-3-030-21496-8_1

If simulating a device by solving an analysis problem gives a fundamental information about its behavior, the design of MEMS is the main computational challenge which arises nowadays, in particular, the automated design optimization. In fact, the industrial designers are more and more involved in solving synthesis problems based on procedures of automated optimal design which are, in turn, based on analysis problems. This is the rationale behind Chaps. 10 and 11: in the former, definitions and properties of synthesis (or inverse) problems are summarized and a few regularization methods are presented, while in the latter methods for the automated optimal design are presented and discussed.

Because numerical methods have to follow the ongoing technological trend, which is more and more oriented towards Nano Electro Mechanical Systems (NEMS), in Chap. 12 a selection of NEMS devices is presented; accordingly, it is shown how they can be still modelled by means of numerical methods used for MEMS.

A categorization commonly accepted of MEMS devices is based on the inherent actuation principle. In fact, there are many principles of actuation, among the others electrical, magnetic, thermal, fluidic and chemical actuation. Each kind of actuation presents advantages and drawbacks; in this book, MEMS devices characterized by electrical, thermal and magnetic actuation are treated.

Electrical actuation is the most common and the oldest one. In fact, capacitive transduction, coupled with the electrostatic actuation, is used for many applications like e.g. pressure sensors, micromotors, accelerometers, gyroscopes and energy harvesters. In particular, pressure sensors became the first mass-produced MEMS device around 1995. On the other hand, electrostatic micromotors were the first devices to be designed and prototyped, exploiting the Silicon integrated technology as early as the late eighties of last century.

In the book the thermal actuation is presented as coupled with the electrical actuation, i.e. the conduction current heats a structural component of the device thanks to the Joule effect; then, the gradient of temperature gives rise to a strain and, finally, to a displacement. This kind of electro-thermo-elastic device is significant from the point of view of complex models, because of the three coupled fields.

Finally, magnetic actuation is considered due to its many advantages. Among the others, low voltages are needed for power supply and, hence, the power consumption is low; moreover, they are simple to control because of their linear response to the input signal. Even if they appear not to be used very extensively, they fill some important niches in mechatronics e.g. for those applications that need large force densities and broad strokes. A more physical insight on electric and magnetic actuations is presented in Chap. 7.

Following the three actuation types, in Chap. 13, examples of MEMS devices like micromotors, accelerometers, micromirrors and actuators of various shapes are analyzed by means of field models. In turn, in Chap. 14 the same devices are considered from the viewpoint of automated optimal design and solved by means of recently proposed optimization algorithms. This part of the book delves into emerging research topics.

Because of the multidisciplinary nature of the covered topics, the skills of different authors were exploited in preparing the book; in this respect, the two main authors gratefully acknowledge the contributions of Dr. Maria Evelina Mognaschi with the University of Pavia in preparing Chaps. 11, 13 and 14 and Prof. Teodor Gotszalk with the Wrocław University of Science and Technology in preparing Chap. 12.

The book is mainly recommended and addressed to senior undergraduate and graduate students of departments of electrical and computer engineering; however, it could also be helpful for students preparing their Ph.D. projects, as well as for researchers and engineers working in the broad area of mechatronics.

Pavia, Łódź
March 2019

Paolo Di Barba, Sławomir Wiak

Chapter 2
MEMS Modelling: Distributed versus Lumped Parameter Models

The miniaturisation of electromechanical systems is impacting our society as deeply as did the mass production of electronic systems in the last few decades. However, only in more recent times the design of MEMS has been approached in a systematic way, by employing methods and algorithms of automated optimal design (AOD).

Accordingly, the design problem is set up as a problem of non-linear optimization of a design criterion (objective function) simultaneously dependent on a vector of design variables and subject to a set of constraints. A generalized formulation of the objective function is in terms of the norm of the discrepancy between prescribed and actual behaviour of the device; this might give rise to a non-trivial mathematical problem to solve at least in an approximate way. The approach implies suitable computational environments made available by the progress in artificial intelligence, where modelling tools are integrated with soft computing tools.

In Fig. 2.1 a principle flow-chart, showing a general procedure of AOD of a MEMS device, is proposed: its core is the interaction between a routine for simulating the device, which is integrated in a loop with a routine minimizing the objective function.

For the sake of generality, in the flow-chart constraints have been categorized as linear ones, which depend on geometrical parameters (for instance admissible ranges of values) and non-linear ones, which depend on physical parameters like e.g. behavioural limits of the device.

The interaction between simulation routine and optimization routine is a rule for any numerical procedure of AOD to work, while the choice of the two relevant algorithms makes the implementation peculiar.

More specifically, the computation of objectives and constraints depends on the model simulating the MEMS device. Here, two main streamlines can be followed, namely distributed parameters or lumped parameters.

Distributed-parameter models—or field models—are:

- one-, two- or three-dimensional, i.e. they take into account geometrical shapes with different degree of accuracy;
- not easy to code because the knowledge of field theory is very specialistic;

P. Di Barba and S. Wiak, *MEMS: Field Models and Optimal Design*, Lecture Notes in Electrical Engineering 573, https://doi.org/10.1007/978-3-030-21496-8_2

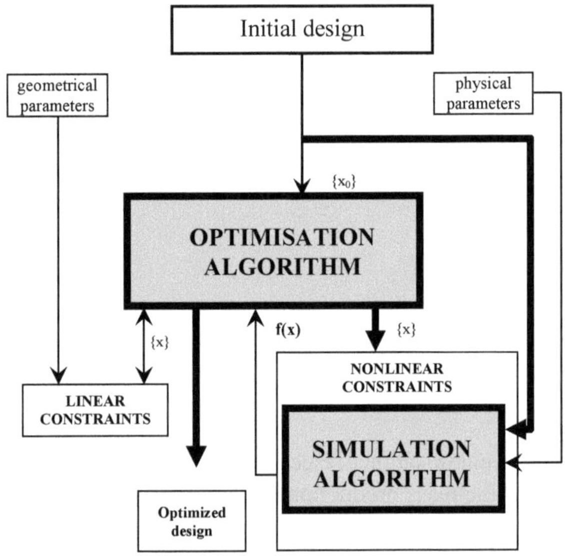

Fig. 2.1 Principle flow-chart of AOD (x, design vector; f(x), objective function)

- not very adequate for fast computing, especially when time-consuming computations in 3D domains take place;
- extremely good for optimal shape design;
- very good for estimating integral parameters (like e.g. capacitance and inductance) of circuit models (field-circuit approach).

The relevant mathematical model is based on partial differential equations belonging either to the parabolic class, if the device behaves in dynamic conditions, or to the elliptic class, if steady-state conditions apply. In electromagnetics, Maxwell's equations equipped with constitutive relationships of materials are the reference; the numerical counterpart is e.g. the finite-element method.

In turn, lumped-parameter models—or circuit models—are:

- zero-dimensional, i.e. space effects are disregarded;
- relatively easy to code because the knowledge of circuit theory is widespread;
- adequate for fast computing;
- unadequate for optimal shape design;
- very good for a preliminary sizing of the device, the design of which could eventually be optimized based on a field model.

The relevant mathematical model is based on ordinary differential equations, if the device behaves in dynamic conditions, or algebraic equations, if steady-state conditions apply. In electrical systems, Kirchhoff's equations equipped with Ohm's equations are the reference; the numerical counterpart are e.g. tableau analysis methods.

Analysis by means of field models is somehow opposed to the traditional analysis based on lumped parameters (i.e. circuit models); in fact, several papers and books based on circuit models can be found in the literature [1]. Put simply, circuit models describe the device behavior as a black box; in turn, field models simulate the internal scenario, more often than not a multiphysics one, which characterizes most of the MEMS devices commonly used.

All in one, the choice between the two categories of models depend on a cost-accuracy trade off. In the following, models stemming from both categories will be used; major emphasis will be devoted to field models whenever space-dependent effects have to be considered.

Reference

1. Jones TB, Nenadic NG (2013) Electromechanics and MEMS. Cambridge University Press

Chapter 3
Engineering Electrostatics and Boundary-Value Problems

3.1 Constitutive Law of Dielectric Material

In a domain Ω with boundary Γ, filled in by a dielectric material, in the presence of free electric charges distributed with density ρ (C m^{-3}) in Ω and/or electric charges distributed with density σ (C m^{-2}) along Γ, the electrostatic field is defined by field strength $\overline{E}$(V m^{-1}) as well as by flux density $\overline{D}$(C m^{-2}). The two vectors are linked by the constitutive law, which, if the material is isotropic and linear, and in the absence of permanent polarization, is

$$\overline{D} = \varepsilon \overline{E} \tag{3.1}$$

The parameter ε is called permittivity (F m^{-1}) and can be factorized as $\varepsilon = \varepsilon_0 \varepsilon_r$ where $\varepsilon_0 = (36\pi)^{-1} 10^{-9}$ F m^{-1} is the permittivity of the empty space, while ε_r is the relative permittivity. The observer is supposed to be at rest with respect to the field [4].

3.2 Maxwell's Equations of Electrostatic Field

The electrostatic field is governed by the following equations in Ω

$$\overline{\nabla} \times \overline{E} = 0 \tag{3.2}$$

$$\overline{\nabla} \cdot \overline{D} = \rho \tag{3.3}$$

and along Γ

$$\bar{n} \times \overline{E} = 0 \tag{3.4}$$

if Γ is a perfect conductor (flux lines perpendicular to Γ), or

P. Di Barba and S. Wiak, *MEMS: Field Models and Optimal Design*, Lecture Notes in Electrical Engineering 573, https://doi.org/10.1007/978-3-030-21496-8_3

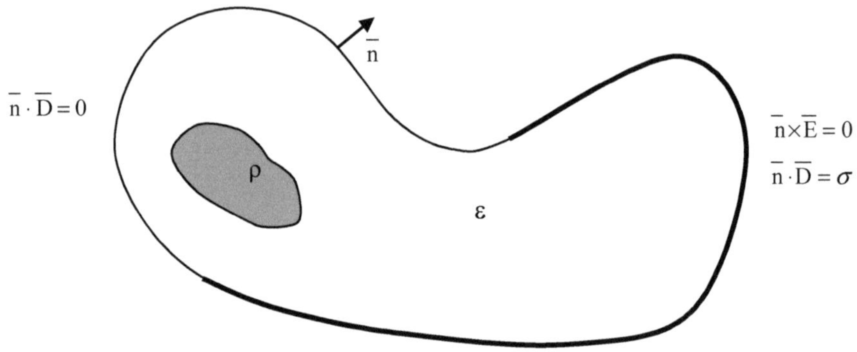

Fig. 3.1 Electric field domain with boundary conditions

$$\bar{n} \cdot \overline{D} = \sigma \tag{3.5}$$

if Γ carries charge density σ (C m^{-2}), or

$$\bar{n} \cdot \overline{D} = 0 \tag{3.6}$$

if Γ is a perfect insulator (flux lines parallel to Γ). A general field domain is shown in Fig. 3.1.

In terms of just vector $\overline{E}$, the equations governing the electrostatic field for a homogeneous, isotropic and linear material become in Ω

$$\overline{\nabla} \times \overline{E} = 0 \tag{3.7}$$

$$\overline{\nabla} \cdot \overline{E} = \frac{\rho}{\varepsilon} \tag{3.8}$$

and along Γ

$$\bar{n} \times \overline{E} = 0 \tag{3.9}$$

or

$$\bar{n} \cdot \overline{E} = \frac{\sigma}{\varepsilon} \quad \text{(for } \Gamma \text{ conducting)} \tag{3.10}$$

and

$$\bar{n} \cdot \overline{E} = 0 \quad \text{(for } \Gamma \text{ insulating)} \tag{3.11}$$

respectively.

According to the Helmholtz's theorem (see Appendix), $\overline{E}$ is uniquely defined and, because of (3.2), it is irrotational.

Moreover, if both σ and ρ are given, then according to (3.3), (3.5) and Gauss's theorem (see A.10) it must be

$$\int_{\Gamma} |\sigma| d\Gamma = \int_{\Omega} |\rho| d\Omega \tag{3.12}$$

i.e. the total charge sums up to zero; in other words, surface and volume charge densities are not independent.

In a two-dimensional domain, at the interface between two different materials of permittivity ε_1 and ε_2, respectively (Fig. 3.2), in the presence of free charge of density σ (C m^{-2}) at the interface, from (A.10), integrating $\overline{D}$ along a closed surface including point P, the following transmission conditions holds

$$\bar{n}_1 \cdot \overline{D}_2 + \bar{n}_2 \cdot \overline{D}_1 = \bar{n}_1 \cdot \overline{D}_2 - \bar{n}_1 \cdot \overline{D}_1 = \bar{n}_1 \cdot \left(\overline{D}_2 - \overline{D}_1\right) = \sigma \tag{3.13}$$

If $\sigma = 0$, the normal component of $\overline{D}$ is continuous.

Then, from (A.11), integrating $\overline{E}$ along a closed rectangular line surrounding point P, it turns out to be

$$\bar{t}_1 \cdot \overline{E}_1 + \bar{t}_2 \cdot \overline{E}_2 = \bar{t}_1 \cdot \overline{E}_1 - \bar{t}_1 \cdot \overline{E}_2 = \bar{t}_1 \cdot \left(\overline{E}_1 - \overline{E}_2\right) = 0 \tag{3.14}$$

i.e. the tangential component of $\overline{E}$ is continuous. The integration line is supposed to have the two smaller sides of vanishing length, and the two larger sides parallel to the tangential unit vectors in P.

In the case of a non-homogeneous material, the following remark can be put forward. After (3.1) and (3.2), considering vector identity (A.16), it turns out to be

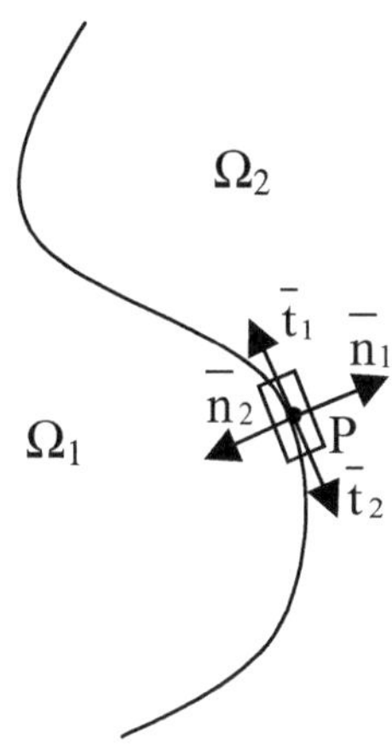

Fig. 3.2 Interface between two field regions Ω_1 and Ω_2

$$\overline{\nabla} \times \varepsilon^{-1}\overline{D} = \varepsilon^{-1}\overline{\nabla} \times \overline{D} + \overline{\nabla}\varepsilon^{-1} \times \overline{D} = 0 \tag{3.15}$$

Apparently, field $\overline{D}$ is irrotational if $\overline{\nabla}\varepsilon^{-1}$ and $\overline{D}$ are parallel vectors; this means that lines separating layers of different ε are orthogonal to field lines of $\overline{D}$. If $\overline{\nabla}\varepsilon^{-1} = 0$ (homogeneous material), then $\overline{D}$ is always irrotational.

Conversely, in a charge-free domain, after (3.1) and (3.3), considering vector identity (A.14), it follows

$$\overline{\nabla} \cdot \varepsilon\overline{E} = \varepsilon\overline{\nabla} \cdot \overline{E} + \overline{\nabla}\varepsilon \cdot \overline{E} = 0 \tag{3.16}$$

In the case of a non-homogeneous medium, field $\overline{E}$ is solenoidal if $\overline{\nabla}\varepsilon$ and $\overline{E}$ are orthogonal vectors; this means that lines separating layers of different ε are parallel to field lines of $\overline{E}$. If $\overline{\nabla}\varepsilon = 0$ (homogeneous material), then $\overline{E}$ is always solenoidal [4].

Finally, an extension of constitutive law (3.1) is considered.

When a permanent polarization $\overline{D}_0$ is present in the dielectric material (electret), the constitutive law becomes

$$\overline{D} = \varepsilon\overline{E} + \overline{D}_0 \tag{3.17}$$

Then, the field equations are

$$\overline{\nabla} \times \overline{E} = 0 \tag{3.18}$$

and

$$\overline{\nabla} \cdot \left(\varepsilon\overline{E}\right) = \rho - \overline{\nabla} \cdot \overline{D}_0 \tag{3.19}$$

in terms of $\overline{E}$.

3.3 From Field to Potentials

(i) If Ω is simply-connected, because of (3.2) it is always possible to introduce a scalar function U (potential (V)) defined as

$$\overline{E} = -\overline{\nabla}U \tag{3.20}$$

This way, the field is oriented from higher to lower values of potential. Therefore, (3.3) becomes

$$\overline{\nabla} \cdot \left(\varepsilon\overline{\nabla}U\right) = -\rho \tag{3.21}$$

This is the Poisson's equation governing potential U, which reduces to

$$\nabla^2 U = -\frac{\rho}{\varepsilon} \tag{3.22}$$

for a homogeneous, isotropic and linear domain Ω.

In the case of $\rho = 0$ in Ω, (3.22) is called Laplace's equation and the potential function U fulfilling it is said to be harmonic.

It can be remarked that, adding any constant k to U, all the equations defined above are fulfilled as well; in order to have U uniquely defined, boundary conditions must be added.

In a two-dimensional domain, along the boundary Γ with normal unit vector $\bar{n} = (n_x, n_y)$ and tangential unit vector $\bar{t} = (t_x, t_y) = (-n_y, n_x)$, the condition (3.11) in terms of field $\overline{E} = (E_x, E_y)$ becomes, in terms of potential U,

$$\bar{n} \cdot \overline{E} = n_x E_x + n_y E_y = -n_x \frac{\partial U}{\partial x} - n_y \frac{\partial U}{\partial y} = -\bar{n} \cdot \overline{\nabla} U = -\frac{\partial U}{\partial n} = 0 \tag{3.23}$$

(homogeneous Neumann's condition).

Similarly, condition (3.9) becomes

$$\begin{aligned} \bar{n} \times \overline{E} &= \bar{i}_z (n_x E_y - n_y E_x) = \bar{i}_z \left(t_y \frac{\partial U}{\partial y} + t_x \frac{\partial U}{\partial x} \right) \\ &= \bar{i}_z (\bar{t} \cdot \overline{\nabla} U) = \bar{i}_z \frac{\partial U}{\partial t} = 0 \end{aligned} \tag{3.24}$$

In the latter case U is constant along Γ (Dirichlet's condition). Lines of equal U are called equipotential lines.

A boundary-value problem is one in which (3.21) is the governing equation, subject to known boundary conditions which may be (3.23) (Neumann's problem) or (3.24) (Dirichlet's problem) or, more generally, (3.23) and (3.24) along Γ_1 and Γ_2, respectively, with $\Gamma = \Gamma_1 \cup \Gamma_2$ and $0 = \Gamma_1 \cap \Gamma_2$.

(ii) If $\rho = 0$ in Ω, then, along with scalar potential U, a vector potential $\overline{A}$ (flux, stream (C m^{-1})) can be uniquely defined, specifying its curl

$$\overline{D} = \overline{\nabla} \times \overline{A} \tag{3.25}$$

and its divergence (gauge condition)

$$\overline{\nabla} \cdot \overline{A} = 0 \tag{3.26}$$

In fact, since $\overline{\nabla} \cdot (\overline{\nabla} \times \overline{A}) = 0$ holds, (3.3) is always fulfilled, while (3.2) becomes

$$\overline{\nabla} \times \varepsilon^{-1} (\overline{\nabla} \times \overline{A}) = 0 \tag{3.27}$$

The latter is Laplace's vector equation governing flux $\overline{A}$. It can be remarked that the gradient of an harmonic function η may be added to $\overline{A}$, having all the above equations fulfilled. In fact, if $\overline{\nabla} \times \overline{A} = 0$ holds, then

$$\overline{A} = -\overline{\nabla}\eta \tag{3.28}$$

and, due to (3.26), it results

$$-\overline{\nabla} \cdot \overline{A} = \overline{\nabla} \cdot \overline{\nabla}\eta = \nabla^2\eta = 0 \tag{3.29}$$

In view to have $\overline{A}$ defined uniquely, suitable boundary conditions along Γ and, if necessary, cuts in Ω must be introduced, transforming a multiply-connected domain into a simply-connected one (Fig. 3.3).

For a homogeneous domain, Laplace's equation (3.27) becomes

$$\overline{\nabla} \times \left(\overline{\nabla} \times \overline{A}\right) = 0 \tag{3.30}$$

and therefore

$$\overline{\nabla}\left(\overline{\nabla} \cdot \overline{A}\right) - \overline{\nabla}^2\overline{A} = 0 \tag{3.31}$$

or, because of (3.26)

$$-\overline{\nabla}^2\overline{A} = 0 \tag{3.32}$$

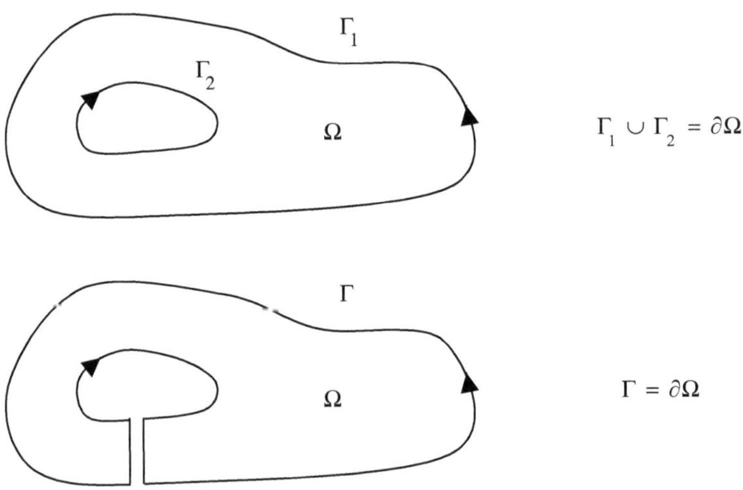

Fig. 3.3 Transforming a doubly-connected domain into a simply-connected one

(iii) When a permanent polarization $\overline{D}_0$ is present (electret), from (3.19) and (3.20) the field equation in terms of scalar potential U is

$$\overline{\nabla} \cdot \left(\varepsilon \overline{\nabla} U\right) = -\rho + \overline{\nabla} \cdot \overline{D}_0 \tag{3.33}$$

3.3.1 Field of a Point Charge in a Spherical Domain

A point charge q (C), located at the origin of a system of spherical coordinates (r, θ, φ), is considered [1].

The symmetry implies $\overline{D} = (D, 0, 0)$; after (A.19), the field equation is

$$\overline{\nabla} \cdot \overline{D} = r^{-2} \frac{\partial \left(r^2 D\right)}{\partial r} = \frac{\partial D}{\partial r} + \frac{2}{r} D = q\delta(r), \quad r > 0 \tag{3.34}$$

where D vanishes as r approaches infinity. Treating (3.34) as a first-order differential equation in terms of $r^2 D$, one has

$$r^2 D(r) = \int_0^r q\delta(\rho)\rho^2 d\rho + k \tag{3.35}$$

where k is the integration constant; then, the general solution is:

$$D(r) = r^{-2} \left(q \int_0^r \rho^2 \delta(\rho) d\rho + k \right) \tag{3.36}$$

The Dirac's δ in a spherical geometry can be approximated by

$$\delta = \lim_{\alpha \to 0} \delta_\alpha \quad \alpha > 0 \tag{3.37}$$

with $\delta_\alpha = \frac{3}{4\pi\alpha^3}$ if $\rho \leq \alpha$ and $\delta_\alpha = 0$ elsewhere. Consequently, field D can be approximated as

$$D = \lim_{\alpha \to 0} D_\alpha \tag{3.38}$$

For $r \leq \alpha$ one has

$$D_\alpha = \frac{1}{r^2} \left(q \int_0^r \rho^2 \delta_\alpha d\rho + k_\alpha \right) = \frac{1}{r^2} \left(\frac{3q}{4\pi\alpha^3} \frac{r^3}{3} + k_\alpha \right) \tag{3.39}$$

namely

$$D_\alpha = \frac{qr}{4\pi\alpha^3} + \frac{k_\alpha}{r^2} \tag{3.40}$$

Since δ_α is a regular function near the origin, also D_α will be regular near zero; therefore $k_\alpha = 0$.

For $r \geq \alpha$ one has

$$\begin{aligned} D_\alpha &= \frac{1}{r^2}\left(q\int_0^\alpha \rho^2\delta_\alpha d\rho + k_\alpha\right) = \frac{1}{r^2}\frac{3q}{4\pi\alpha^3}\int_0^\alpha \rho^2 d\rho \\ &= \frac{1}{r^2}\frac{3q}{4\pi\alpha^3}\frac{\alpha^3}{3} = \frac{q}{4\pi r^2} \end{aligned} \tag{3.41}$$

hence

$$D_\alpha(r) = \frac{qr}{4\pi\alpha^3}, \quad r \leq \alpha \tag{3.42}$$

and

$$D_\alpha(r) = \frac{q}{4\pi r^2}, \quad \alpha < r \tag{3.43}$$

Coulomb's law follows

$$D(r) = \lim_{\alpha\to 0} D_\alpha(r) = \frac{q}{4\pi r^2}, \quad r > 0 \tag{3.44}$$

Finally, the potential results

$$U(r) = -\int_\infty^r \frac{q}{4\pi\varepsilon\rho^2} d\rho = \frac{q}{4\pi\varepsilon r}, \quad r > 0 \tag{3.45}$$

3.3.2 *Field of a Point Charge Surrounded by a Spherical Surface at Known Potential*

The source charge $+q$ gives rise to an induced charge $-q$ on the sphere; in addition to it, the free charge on the sphere, originating the potential $u = k$, is $q' = 4\pi\varepsilon Rk$ (see 3.45). The total charge on the sphere is $q_t = 4\pi\varepsilon Rk - q$.

From Gauss's theorem the field turns out to be

$$E(r) = \frac{q}{4\pi \varepsilon r^2}, \quad 0 < r < R; \;\; E(r) = \frac{Rk}{r^2}, \quad r > R \tag{3.46}$$

The potential is given by

$$U(r) = q\left(\frac{1}{4\pi r} - \frac{1}{4\pi R}\right) + k, \quad 0 < r < R \tag{3.47}$$

$$U(r) = \frac{q'}{4\pi r} = k\frac{R}{r}, \quad r > R \tag{3.48}$$

It can be assessed that $U(r)$ solves the particular case of $q = 0$, for which, after (3.21) in spherical coordinates

$$\nabla^2 U(r) = U''(r) + \frac{2}{r}U'(r) = 0, \quad U(R) = k, \;\; U(\infty) = 0 \tag{3.49}$$

holds.

Potential and field strength are shown in Figs. 3.4 and 3.5, respectively.

At $r = R$, u is continuous for any k, while E is not if $k \neq \frac{q}{4\pi \varepsilon R}$. The particular cases of a grounded sphere and a supplied sphere follow, when $k = 0$ with $q \neq 0$, and $k \neq 0$ with $q = 0$, respectively.

Fig. 3.4 Potential versus position

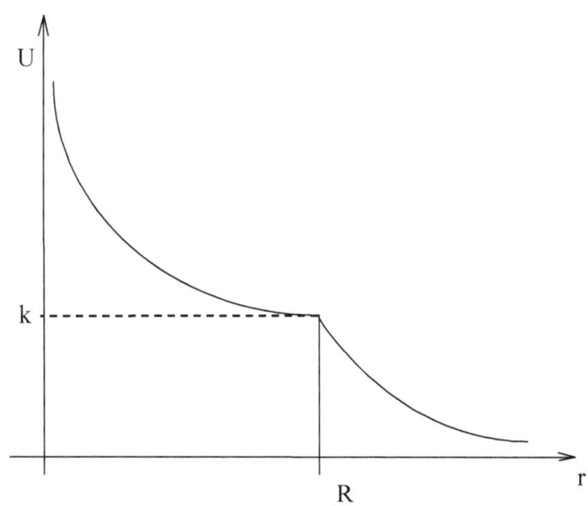

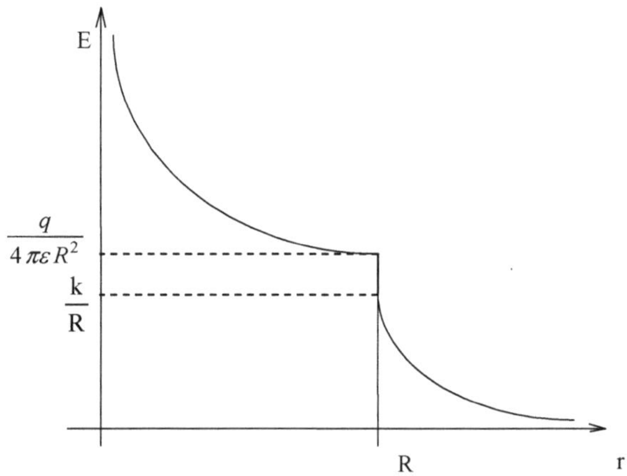

Fig. 3.5 Field versus position ($q > 4\pi\varepsilon Rk$)

3.3.3 Field of a Charged Plane in a Rectangular Domain

A conducting plane (y, z) of infinite extension with a uniform distribution σ (C m^{-2}) of charge is considered [1]. Due to the field symmetry one has $\overline{D} = (D, 0, 0)$ so that the problem can be formulated as:

$$\overline{\nabla} \cdot \overline{D} = \sigma\delta(x) = \frac{\partial(\varepsilon E)}{\partial x} \tag{3.50}$$

where $\delta(x)$ is the delta function (or Dirac's delta) at $x = 0$. By integrating the latter equation along $x \in (-\infty, \infty)$, it follows

$$\frac{\sigma}{\varepsilon} = E(\infty) - E(-\infty) \tag{3.51}$$

Taking into account the field orientation with respect to the plane, the boundary condition

$$E(\infty) = -E(-\infty) \tag{3.52}$$

can be set up. Therefore, it turns out to be

$$E(x) = \frac{\sigma}{2\varepsilon}\text{sgn}(x) \tag{3.53}$$

The potential is

$$U(x) = \frac{\sigma}{2\varepsilon}|x| \tag{3.54}$$

3.4 Electrostatic Energy

Given an electrostatic field characterized by strength $\overline{E}$ and flux density $\overline{D}$ in a linear dielectric material, the specific energy (J m^{-3}) stored in the field is defined as $\frac{1}{2}\overline{E}\cdot\overline{D}$.

Taking into account (A.14), one has

$$\overline{E}\cdot\overline{D} = -\overline{\nabla}U\cdot\overline{D} = U\overline{\nabla}\cdot\overline{D} - \overline{\nabla}\cdot\left(U\overline{D}\right) \tag{3.55}$$

Then, the total energy W (J) stored in a region Ω of boundary Γ, is

$$\begin{aligned} W &= \frac{1}{2}\int_{\Omega}\overline{E}\cdot\overline{D}d\Omega = \frac{1}{2}\int_{\Omega}U\rho d\Omega - \frac{1}{2}\int_{\Omega}\overline{\nabla}\cdot\left(U\overline{D}\right)d\Omega \\ &= \frac{1}{2}\int_{\Omega}U\rho d\Omega - \frac{1}{2}\int_{\Gamma}U\overline{D}\cdot\bar{n}d\Gamma \end{aligned} \tag{3.56}$$

The equation above gives an expression of the total energy as a function of potential U in terms of source ρ in the region Ω and source $\sigma = \overline{D}\cdot\bar{n}$ on the boundary Γ.

If the material is isotropic, the energy W (J) stored in region Ω is given by:

$$W = \frac{1}{2}\int_{\Omega}EDd\Omega \tag{3.57}$$

Under the same assumptions, if the constitutive relationship of the dielectric material is non-linear, the specific energy is $\int_0^D EdD'$ and the total energy W is

$$W = \int_{\Omega}\left(\int_0^D EdD'\right)d\Omega \tag{3.58}$$

In some cases it is convenient to introduce the specific co-energy $\int_0^E DdE'$ so that the total co-energy W' is

$$W' = \int_{\Omega}\left(\int_0^E DdE'\right)d\Omega \tag{3.59}$$

In the case of a linear medium $W = W'$ results.

3.5 Forces and Torques in the Electrostatic Field

In order to evaluate the mechanical effect on structures located in the field region, three methods can be applied [2].

3.5.1 Principle of Virtual Work

Given the structure, on the gravity centre of which force $\overline{F}$ is to be calculated, a virtual linear displacement *ds* in the direction of $\overline{F}$, supposing that the field source $q = \int_{\Omega} \rho d\Omega$ is constant, mechanical work *Fds* and variation of internal energy *dW* take place so that the energy balance

$$Fds + dW = 0 \tag{3.60}$$

holds.

Therefore, the force experienced can be evaluated as

$$F = -\frac{dW}{ds} \tag{3.61}$$

with *W* given by (3.58).

Similarly, in the case of a virtual angular displacement $d\vartheta$, the torque *M* with respect to the rotation axis is

$$M = -\frac{dW}{d\vartheta} \tag{3.62}$$

The system spontaneously tends to assume the configuration corresponding to the minimum energy stored.

On the other hand, if the structure has constant voltage *U*, the energy balance is

$$Fds + dW = Udq \tag{3.63}$$

$$Fds = d(qU - W) \tag{3.64}$$

$$F = \frac{d}{ds}(qU - W) \tag{3.65}$$

where *Udq* is the electric work done to keep *U* constant.

Correspondingly, the torque is

$$M = \frac{d}{d\vartheta}(qU - W) \tag{3.66}$$

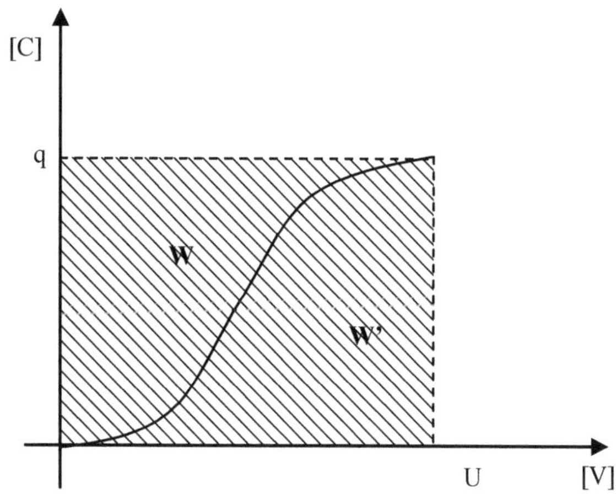

Fig. 3.6 Electrostatic energy W and co-energy W′ in the charge-potential plane

The quantity $qU - W$, denoted by W', is called complementary energy or co-energy, given by (3.59), of the system; if the system is linear, W' and W coincide.

If U is kept constant, the system evolution is towards the geometry of maximum co-energy.

The geometric interpretation of energy and co-energy in the charge-potential plane is straightforward (Fig. 3.6).

There are two possible ways, in fact, to move the system from the initial point (0, 0) to the final point (U, q), namely:

(i) from 0 to q by means of increments of free charge on the electrodes; in this case, the system is electrically insulated and the total work done is W (energy);
(ii) from 0 to U by, by means of increments of voltage between the electrodes; in this case, the system is electrically connected to a voltage source and the total work done is W' (co-energy).

It is always $W + W' = qU$, while, if the system is linear, $W = W' = \frac{1}{2}qU$ holds.

3.5.2 Coulomb's Method

It is based on the definition of electric field; the force $\overline{F}$ exerted on a point charge q in the empty space is $\overline{F} = q\overline{E}$ where $\overline{E}$ is the external field, i.e. the field in the absence of charge q.

If free charge q, distributed with density ρ in a region Ω exhibiting permittivity ε, is considered, the force turns out to be

$$\overline{F} = \int_{\Omega} \rho \overline{E} d\Omega = \int_{\Omega} (\overline{\nabla} \cdot \overline{D}) \overline{E} d\Omega \qquad (3.67)$$

Directions of force and electric field are coincident.

3.5.3 Method of Maxwell's Stress Tensor

Having defined a closed surface Γ enclosing the structure, then force $\overline{F}$ is evaluated as

$$\overline{F} = \int_{\Omega} \nabla \cdot \overline{\overline{T}} d\Omega = \int_{\Gamma} \overline{\overline{T}} \cdot \bar{n} d\Gamma \tag{3.68}$$

where $\bar{n}$ is the outward normal unit vector and $\overline{\overline{T}}$ is defined as the Maxwell's electric stress tensor. It can be represented as a matrix whose elements are specific energies related to the field; assuming a system of rectangular coordinates, in a three-dimensional domain, tensor $\overline{\overline{T}}$ is

$$\overline{\overline{T}} = \begin{bmatrix} \frac{1}{2}(E_x D_x - E_y D_y - E_z D_z) & E_x D_y & E_x D_z \\ E_y D_x & \frac{1}{2}(E_y D_y - E_x D_x - E_z D_z) & E_y D_z \\ E_z D_x & E_z D_y & \frac{1}{2}(E_z D_z - E_x D_x - E_y D_y) \end{bmatrix} \tag{3.69}$$

For the tensor to be properly defined, the integration surface must not be coincident with the boundary between materials with different values of permittivity.

Let the problem of deriving the Maxwell's stress tensor (3.69) from the Coulomb's law (3.67) be considered; the volume force density is given by:

$$\overline{f} = \rho \overline{E} = (\nabla \cdot \overline{D}) \overline{E} \tag{3.70}$$

In a system of rectangular coordinates, the x-directed component of force density is

$$f_x = \varepsilon^{-1} D_x \frac{\partial D_x}{\partial x} + \varepsilon^{-1} D_y \frac{\partial D_y}{\partial y} + \varepsilon^{-1} D_z \frac{\partial D_z}{\partial z} \tag{3.71}$$

The latter can be rewritten as

$$\begin{aligned} f_x = & \frac{1}{2} \varepsilon^{-1} \frac{\partial D_x^2}{\partial x} + \varepsilon^{-1} \frac{\partial (D_x D_y)}{\partial y} - \varepsilon^{-1} D_y \frac{\partial D_x}{\partial y} \\ & + \varepsilon^{-1} \frac{\partial (D_x D_z)}{\partial z} - \varepsilon^{-1} D_z \frac{\partial D_x}{\partial z} \end{aligned} \tag{3.72}$$

and also as

$$f_x = \frac{1}{2}\varepsilon^{-1}\frac{\partial\left(D_x^2 - D_y^2 - D_z^2\right)}{\partial x} + \varepsilon^{-1}\frac{\partial\left(D_x D_y\right)}{\partial y} + \varepsilon^{-1}\frac{\partial(D_x D_z)}{\partial z}$$
$$+ \frac{1}{2}\varepsilon^{-1}\frac{\partial D_y^2}{\partial x} + \frac{1}{2}\varepsilon^{-1}\frac{\partial D_z^2}{\partial x} - \varepsilon^{-1}D_y\frac{\partial D_x}{\partial y} - \varepsilon^{-1}D_z\frac{\partial D_x}{\partial z} \tag{3.73}$$

It follows

$$f_x = \frac{1}{2}\frac{\partial\left(E_x D_x - E_y D_y - E_z D_z\right)}{\partial x} + \frac{\partial\left(E_x D_y\right)}{\partial y} + \frac{\partial(E_x D_z)}{\partial z}$$
$$+ \varepsilon^{-1}D_y\left(\frac{\partial D_y}{\partial x} - \frac{\partial D_x}{\partial y}\right) + \varepsilon^{-1}D_z\left(\frac{\partial D_z}{\partial x} - \frac{\partial D_x}{\partial z}\right) \tag{3.74}$$

Due to the assumption of irrotational field (3.2), the fourth and the fifth terms of the right-hand side of (3.74) are zero. Therefore, f_x can be expressed as the divergence of a primitive vector $\bar{v}_1$

$$f_x = \overline{\nabla} \cdot \bar{v}_1 \tag{3.75}$$

with

$$\bar{v}_1 = \left(\frac{1}{2}\left(E_x D_x - E_y D_y - E_z D_z\right), E_x D_y, E_x D_z\right) \tag{3.76}$$

apart from a constant that can be set to zero.

In a similar way, it can be proven that

$$f_y = \overline{\nabla} \cdot \bar{v}_2 \tag{3.77}$$

with

$$\bar{v}_2 = \left(E_y D_x, \frac{1}{2}\left(E_y D_y - E_x D_x - E_z D_z\right), E_y D_z\right) \tag{3.78}$$

and

$$f_z = \overline{\nabla} \cdot \bar{v}_3 \tag{3.79}$$

with

$$\bar{v}_3 = \left(E_z D_x, E_z D_z, \frac{1}{2}\left(E_z D_z - E_x D_x - E_y D_y\right)\right) \tag{3.80}$$

The components of vectors (3.75), (3.77) and (3.79) are just the row entries of matrix (3.69) representing the electric stress tensor. Therefore, according to

(3.68), the force $\overline{F}$ (N) can be computed as the flux, leaving surface Γ, of tensor $\overline{\overline{T}}$ represented by matrix (3.69). Correspondingly, the torque is given by $\overline{M} = \int_\Gamma \bar{r}_{PO} \times \overline{\overline{T}} \cdot \bar{n} d\Gamma$ where $\bar{r}_{PO}$ is the position vector of point P on Γ with respect to the rotation axis in O.

It can be remarked that a solenoidal vector $\bar{w}$ may be added to (3.76), (3.78) and (3.80) leaving force density components (3.75), (3.77) and (3.79) fulfilled. This means that stress tensor (3.69) is not uniquely defined.

As far as a comparison of methods is concerned, the following remark can be put forward. In order Coulomb's method to apply, a charge density must be defined in Ω; in contrast, virtual work principle (VWP) and Maxwell's stress tensor method (MST) are more general. VWP is computationally more expensive, because the derivative of energy or co-energy is approximated by means of a finite difference, involving two displaced positions of the structure. Therefore, two field analyses are necessary to compute force or torque at a given position. MST require only one field analysis.

3.6 Worked Examples

3.6.1 Force Between the Plates of a Plane Capacitor

Let a capacitor exhibiting a pair of parallel plates (electrodes) of infinite extension be considered [3]; the plates are assumed to carry a surface charge density equal to σ and $-\sigma$, respectively (Fig. 3.7).

Knowing the field of a single charged plate, after the principle of superposition it turns out to be from (3.53):

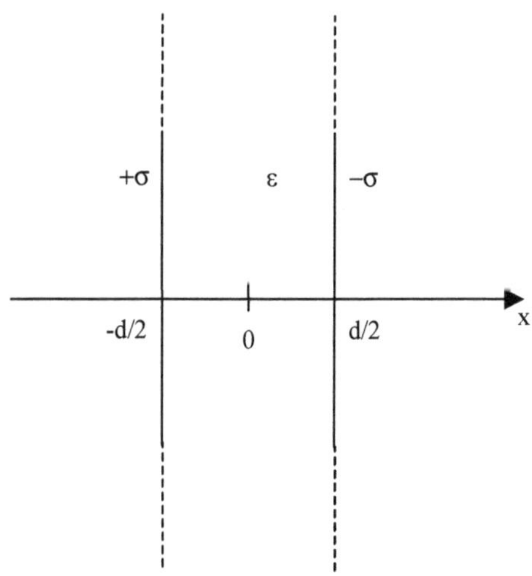

Fig. 3.7 Single-layer plane capacitor

$$x < -\frac{d}{2}, \quad E = -\frac{\sigma}{2\varepsilon} + \frac{\sigma}{2\varepsilon} = 0, \quad U = 0 \tag{3.81}$$

$$-\frac{d}{2} < x < \frac{d}{2}, \quad E = 2\frac{\sigma}{2\varepsilon} = \frac{\sigma}{\varepsilon}, \quad U = -\frac{\sigma}{\varepsilon}\left(x + \frac{d}{2}\right) \tag{3.82}$$

$$x > \frac{d}{2}, \quad E = \frac{\sigma}{2\varepsilon} - \frac{\sigma}{2\varepsilon} = 0, \quad U = -\frac{\sigma}{\varepsilon}d \tag{3.83}$$

If a subsection of finite area S is considered, the relevant capacitance C is defined as:

$$C = \frac{q}{U} = \varepsilon\frac{S}{d} \tag{3.84}$$

Because of Gauss's theorem, $q = \sigma S$ is the charge on each plate of the capacitor.

If the capacitor is subject to a constant voltage U, the co-energy is $W' = \frac{1}{2}CU^2$ and the force acting on a plate is from (3.66)

$$F = \frac{\partial W'}{\partial d} = \frac{1}{2}U^2\frac{\partial C}{\partial d} = -\frac{1}{2}U^2\frac{\varepsilon S}{d^2} \tag{3.85}$$

If, in turn, the charge is constant, the energy is $W = \frac{q^2}{2C}$ and the force is from (3.61):

$$\begin{aligned} F &= -\frac{\partial W}{\partial d} = \frac{1}{2}q^2\frac{1}{C^2}\frac{\partial C}{\partial d} = -\frac{1}{2}\frac{q^2}{C^2}\frac{\varepsilon S}{d^2} \\ &= -\frac{\sigma^2 S}{2\varepsilon} = -\frac{1}{2}U^2\frac{\varepsilon S}{d^2} \end{aligned} \tag{3.86}$$

In both cases, the negative sign denotes an attractive force.

Resorting to the Coulomb's method, supposing the charge to be constant and uniformly distributed, the field external to the plate located at $x = \frac{d}{2}$ is equal to $\frac{\sigma}{2\varepsilon}$ while the charge carried by the plate itself is equal to $-\sigma S$. Therefore, the force acting on the plate of surface S results $-\frac{\sigma^2 S}{2\varepsilon} = -\frac{1}{2}U^2\frac{\varepsilon S}{d^2}$; the force is attractive.

Finally, on the basis of the Maxwell's stress tensor, considering a parallelepiped surface, with an axis orthogonal to a plate and enclosing a portion S of it, one can easily obtain:

$$\overline{\overline{T}} = \begin{bmatrix} \frac{1}{2}E_x D_x & 0 & 0 \\ 0 & -\frac{1}{2}E_x D_x & 0 \\ 0 & 0 & -\frac{1}{2}E_x D_x \end{bmatrix} \tag{3.87}$$

with $\bar{n} = (-1, 0, 0)$ and then

$$\overline{\overline{T}} \cdot \bar{n} = -\frac{1}{2}\varepsilon E_x^2 \tag{3.88}$$

$$F = -\frac{1}{2}\varepsilon E_x^2 S = -\frac{1}{2}U^2\frac{\varepsilon S}{d^2} \tag{3.89}$$

The force is attractive.

3.6.2 Force at the Interface Between Two Dielectric Materials

Let two layers of dielectric materials characterized by permittivities ε_1 and ε_2, respectively, and subject to an applied voltage V as shown in Fig. 3.8, be considered [1].

Using obvious notations, voltages and field strengths in each layer are such that

$$E_1 x + E_2(h - x) = V_1 + V_2 = V \tag{3.90}$$

and

$$\varepsilon_1 E_1 = \varepsilon_2 E_2 \tag{3.91}$$

following (3.20) in its integral form and (3.13) with $\sigma = 0$, respectively.

Therefore, one has

$$E_1 = \frac{\varepsilon_2 V}{\varepsilon_1(h - x) + \varepsilon_2 x}, \quad E_2 = \frac{\varepsilon_1 V}{\varepsilon_1(h - x) + \varepsilon_2 x} \tag{3.92}$$

According to (3.59), the co-energy of the system is

$$W'(x) = \frac{1}{2}\varepsilon_1 E_1^2 Sx + \frac{1}{2}\varepsilon_2 E_2^2 S(h - x) = \frac{1}{2}\varepsilon_1\varepsilon_2 \frac{V^2 S}{\varepsilon_1(h - x) + \varepsilon_2 x} \tag{3.93}$$

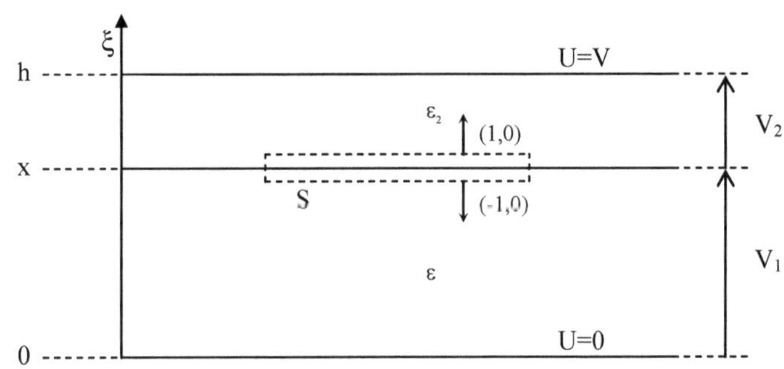

Fig. 3.8 Two-layer plane capacitor

By means of the principle of virtual work (3.65), the force acting at the interface between the two layers turns out to be

$$F = \frac{\partial W'}{\partial x} = \frac{1}{2}(\varepsilon_1 - \varepsilon_2)\frac{\varepsilon_1 \varepsilon_2 V^2 S}{[\varepsilon_1(h-x) + \varepsilon_2 x]^2} \tag{3.94}$$

If $\varepsilon_1 \neq \varepsilon_2$ the dielectric with higher permittivity tends to penetrate into the other one; if $\varepsilon_1 = \varepsilon_2$ the force is zero.

By means of Maxwell's stress tensor, a parallelepiped surface, incorporating the inner boundary between the two dielectric materials and parallel to it, is considered as the integration surface. It results:
inside layer 1

$$\overline{\overline{T}} \cdot \bar{n} = -\frac{1}{2}\varepsilon_1 E_1^2 \tag{3.95}$$

$$F_1 = -\frac{1}{2}\varepsilon_1 E_1^2 S = -\frac{\varepsilon_1 \varepsilon_2^2 V^2 S}{2[\varepsilon_1(h-x) + \varepsilon_2 x]^2} \tag{3.96}$$

inside layer 2

$$\overline{\overline{T}} \cdot \bar{n} = \frac{1}{2}\varepsilon_2 E_2^2 \tag{3.97}$$

$$F_2 = \frac{1}{2}\varepsilon_2 E_2^2 S = \frac{\varepsilon_1^2 \varepsilon_2 V^2 S}{2[\varepsilon_1(h-x) + \varepsilon_2 x]^2} \tag{3.98}$$

Therefore, the total force at the interface between layer 1 and layer 2 is

$$F = F_1 + F_2 = \frac{1}{2}(\varepsilon_1 - \varepsilon_2)\frac{\varepsilon_1 \varepsilon_2 V^2 S}{[\varepsilon_1(h-x) + \varepsilon_2 x]^2} \tag{3.99}$$

Resorting to the Coulomb's method, supposing the charge to be constant and uniformly distributed, the field $\widetilde{E}(h)$ external to the plate located at $\xi = h$ is equal to

$$\widetilde{E}(h) = \frac{1}{2}E_2 = \frac{\varepsilon_1 V}{2[\varepsilon_1(h-x) + \varepsilon_2 x]} \tag{3.100}$$

while the charge carried by the plate itself is equal to

$$q_2 = D_2 S = \varepsilon_2 E_2 S = \frac{\varepsilon_1 \varepsilon_2 V S}{\varepsilon_1(h-x) + \varepsilon_2 x} \tag{3.101}$$

Then, the force acting on the plate of surface S at $\xi = h$ is

$$F_2 = q_2 \widetilde{E}(h) = \frac{\varepsilon_1^2 \varepsilon_2 V^2 S}{2[\varepsilon_1(h-x)+\varepsilon_2 x]^2} \tag{3.102}$$

Conversely, the field $\widetilde{E}(0)$ external to the plate located at $\xi = 0$ is equal to

$$\widetilde{E}(0) = \frac{1}{2} E_1 = \frac{\varepsilon_2 V}{2[\varepsilon_1(h-x)+\varepsilon_2 x]} \tag{3.103}$$

while the charge carried by the plate itself is equal to

$$q_1 = -D_1 S = -\varepsilon_1 E_1 S = -\frac{\varepsilon_1 \varepsilon_2 V S}{\varepsilon_1(h-x)+\varepsilon_2 x} \tag{3.104}$$

Then, the force acting on the plate of surface S at $\xi = 0$ is

$$F_1 = q_1 \widetilde{E}(0) = -\frac{\varepsilon_1 \varepsilon_2^2 V^2 S}{2[\varepsilon_1(h-x)+\varepsilon_2 x]^2} \tag{3.105}$$

Finally, the total force acting on the surface at $\xi = x$ between the two dielectric layers is

$$F = \frac{1}{2}(\varepsilon_1 - \varepsilon_2) \frac{\varepsilon_1 \varepsilon_2 V^2 S}{[\varepsilon_1(h-x)+\varepsilon_2 x]^2} \tag{3.106}$$

3.6.3 *Torque on the Rotating Electrodes of a Capacitor*

Let a capacitor exhibiting a pair of movable electrodes be considered. Both electrodes are shaped as circular sectors of radius R and angular width α_0. The electrodes are parallel and can rotate around an axis orthogonal to them.

Neglecting the fringing field, the field between the electrodes is supposed to be uniform; the surface area of the overlapped electrodes is equal to

$$A' = A \frac{\alpha}{\alpha_0} \tag{3.107}$$

with $0 < \alpha < \alpha_0$ overlapping angle and $A = \alpha_0 R_2$ total area of the electrodes. After (3.84), the capacitance is:

$$C = \frac{\varepsilon A'}{d} = \frac{\varepsilon A}{d} \frac{\alpha}{\alpha_0} \tag{3.108}$$

with d distance between electrodes.

Let the operation at constant charge q be considered first. After (3.82), the field strength is $E = q(\varepsilon_0 A')^{-1}$ and the corresponding energy is

$$W = \frac{q^2}{2C} = \frac{q^2 d}{2\varepsilon A}\frac{\alpha_0}{\alpha} \tag{3.109}$$

By means of the virtual work principle, after (3.62) the torque turns out to be

$$M = -\left.\frac{dW}{d\alpha}\right|_{q=const} = \frac{q^2 d}{2\varepsilon A}\frac{\alpha_0}{\alpha^2} \tag{3.110}$$

It can be noted that the torque is inversely proportional to α_2.

Now, let the operation at constant voltage V be considered. The field strength is $E = Vd^{-1}$ and the corresponding co-energy is

$$W' = \frac{1}{2}CV^2 = \frac{V^2 \varepsilon A}{2d}\frac{\alpha}{\alpha_0} \tag{3.111}$$

By means of the virtual work principle, after (3.66) the torque turns out to be

$$M = \left.\frac{dW'}{d\alpha}\right|_{V=const} = \frac{V^2 \varepsilon A}{2d\alpha_0} \tag{3.112}$$

It can be noted that the torque is independent of α.

Due to linearity of the dielectric material, torques are numerically equal; the electrodes spontaneously tend to fully overlap.

References

1. Di Barba P, Savini A, Wiak S (2008) Field models in electricity and magnetism. Springer, Berlin
2. Hammond P (1971) Applied electromagnetism. Pergamon
3. Silvester PP (1968) Modern electromagnetic fields. Prentice-Hall
4. Simonyi K (1963) Foundations of electrical engineering: fields, networks, waves. Pergamon

Chapter 4
Engineering Magnetostatics and Boundary-Value Problems

4.1 Constitutive Law of Magnetic Material

In a domain Ω, having boundary Γ, containing permanent magnets, i.e. aggregates of magnetic dipoles or, from now on, steady electric current distributed with density $\overline{J}$ (A m^{-2}), a magnetostatic field is set up; it is defined by field strength $\overline{H}$ (A m^{-1}) as well as flux density $\overline{B}$ (Wb m^{-2} = T). In general, the link between $\overline{H}$ and $\overline{B}$, i.e. the constitutive law of the medium, is complicated. Neglecting hysteresis, the law is single-valued and can be expressed, for an isotropic medium in the absence of permanent magnetization, by

$$\overline{B} = \mu \overline{H} \tag{4.1}$$

where μ is called permeability (H m^{-1}) and, in the most general case, is a function of $|\overline{H}|$; the inverse of μ is called reluctivity ν. The observer is supposed to be at rest with respect to the field [4].

4.2 Maxwell's Equations of Magnetostatic Field

The equations governing the magnetic field are in Ω

$$\overline{\nabla} \cdot \overline{B} = 0 \tag{4.2}$$

$$\overline{\nabla} \times \overline{H} = \overline{J} \tag{4.3}$$

and along Γ

$$\overline{n} \cdot \overline{B} = 0 \tag{4.4}$$

P. Di Barba and S. Wiak, *MEMS: Field Models and Optimal Design*, Lecture Notes in Electrical Engineering 573, https://doi.org/10.1007/978-3-030-21496-8_4

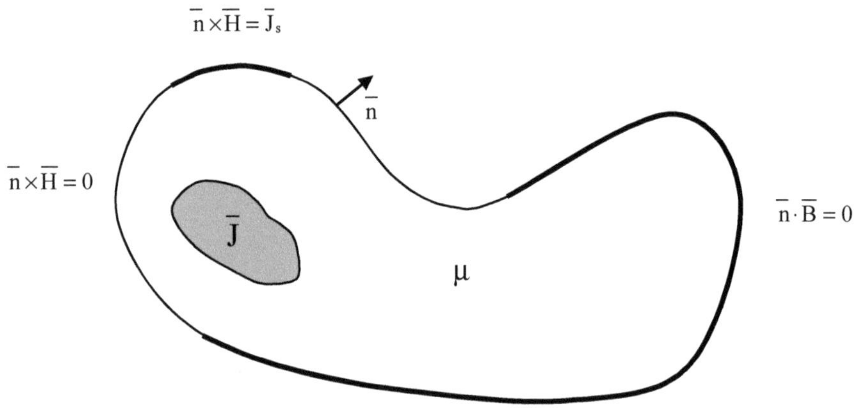

Fig. 4.1 Magnetic field domain with field-based boundary conditions

if Γ is a flux line (flux lines parallel to Γ), or

$$\overline{n} \times \overline{B} = \mu \overline{J}_S \tag{4.5}$$

if current of surface density $\overline{J}_S$ (A m^{-1}) is present, or

$$\overline{n} \times \overline{H} = 0 \tag{4.6}$$

if flux lines are perpendicular to Γ.

For an isotropic and linear medium, in terms of $\overline{B}$, the equations become in Ω

$$\overline{\nabla} \cdot \overline{B} = 0; \quad \overline{\nabla} \times \overline{B} = \mu \overline{J} \tag{4.7}$$

with

$$\overline{n} \cdot \overline{B} = 0 \tag{4.8}$$

or

$$\overline{n} \times \overline{B} = \mu \overline{J}_S \tag{4.9}$$

or

$$\overline{n} \times \overline{H} = 0 \quad \text{along } \Gamma \tag{4.10}$$

The equations written above unambiguously define the magnetostatic field which, because of (4.5), is solenoidal [4].

A general field domain is shown in Fig. 4.1.

Fig. 4.2 Line current

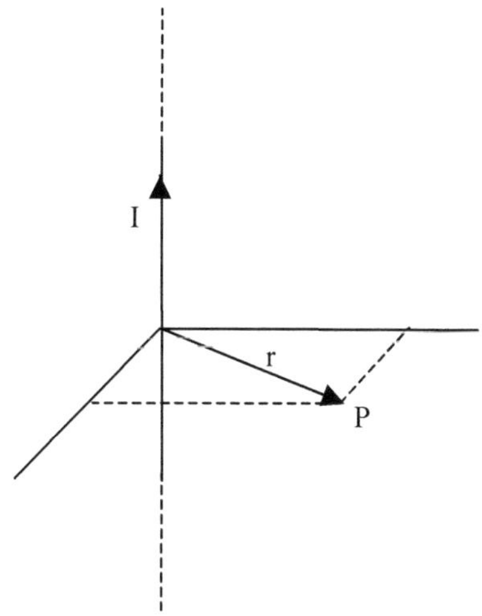

If both $\overline{J}_S$ and $\overline{J}$ are given, then it must be

$$\int_{\Gamma} |\overline{J}_s| d\Gamma = \int_{\Omega} |\overline{J}| d\Omega \tag{4.11}$$

i.e. the total current sums up to zero: therefore, densities J_S and J cannot be independent.

In a non-homogeneous domain at the interface between two materials of permeability μ_1 and μ_2, from (4.2) it holds

$$\overline{n} \cdot \left(\overline{B}_2 - \overline{B}_1\right) = 0 \tag{4.12}$$

so that the normal component of $\overline{B}$ is always continuous (Fig. 4.2).

If there is a current of density $\overline{J}_S$ (A m^{-1}), then from (4.3) it follows

$$\overline{n} \times \left(\overline{H}_2 - \overline{H}_1\right) = \overline{J}_s \tag{4.13}$$

If $\overline{J}_s = 0$ the tangential component of $\overline{H}$ is continuous. Equations (4.12) and (4.13) are called transmission conditions.

In the case of a non-homogeneous medium, the following remark can be put forward. After (4.1) and (4.2), considering vector identity (A.14), one has

$$\overline{\nabla} \cdot \mu\overline{H} = \mu\overline{\nabla} \cdot \overline{H} + \overline{\nabla}\mu \cdot \overline{H} = 0 \tag{4.14}$$

In the case of a non-homogeneous medium, field $\overline{H}$ is solenoidal if $\overline{\nabla}\mu$ and $\overline{H}$ are orthogonal vectors; this means that lines separating layers of different μ are parallel to field lines of $\overline{H}$.

Conversely, after (4.1) and (4.3), considering vector identity (A.16), it turns out to be

$$\overline{\nabla} \times \mu^{-1}\overline{B} = \mu^{-1}\overline{\nabla} \times \overline{B} + \overline{\nabla}\mu^{-1} \times \overline{B} = \overline{J} \tag{4.15}$$

It appears that, in a current-free medium (i.e. J = 0), field $\overline{B}$ is irrotational if $\overline{\nabla}\mu^{-1}$ and $\overline{B}$ are parallel vectors; this means that lines separating layers of different μ are orthogonal to field lines of $\overline{B}$. If $\overline{\nabla}\mu^{-1} = 0$ and J = 0 (homogeneous current-free medium), then $\overline{B}$ is always irrotational.

Finally, an extension of constitutive law (4.1) is considered.

In the presence of a permanent magnetization $\overline{B}_0$ in the magnetic material (permanent magnet) the constitutive law is

$$\overline{B} = \mu\overline{H} + \overline{B}_0 \tag{4.16}$$

In this case the field equations are

$$\overline{\nabla} \cdot \overline{B} = 0 \tag{4.17}$$

$$\overline{\nabla} \times \overline{B} = \mu\overline{J} + \overline{\nabla} \times \overline{B}_0 \tag{4.18}$$

In particular, the field inside a permanent magnet is described by (4.18) with $\overline{J} = 0$; it follows that the magnet can be modelled by an equivalent distribution of current given by $\overline{J}_{eq} = \mu^{-1}\overline{\nabla} \times \overline{B}_0$.

4.3 From Field to Potentials

(i) From (4.2), since, for any vector $\overline{A}$, $\overline{\nabla} \cdot \left(\overline{\nabla} \times \overline{A}\right) = 0$ holds (see A.8), it is possible to define a vector function $\overline{A}$ (Wb m^{-1}) called vector potential by means of

$$\overline{\nabla} \times \overline{A} = \overline{B} \tag{4.19}$$

and

$$\overline{\nabla} \cdot \overline{A} = 0 \quad \text{(gauge condition)} \tag{4.20}$$

This way (4.2) is fulfilled, while (4.3) becomes

$$\overline{\nabla} \times \mu^{-1}\left(\overline{\nabla} \times \overline{A}\right) = \overline{J} \tag{4.21}$$

For a homogeneous domain, after (A.12) and (4.20) it turns out to be

$$\overline{\nabla}^2\overline{A} = -\mu\overline{J} \tag{4.22}$$

This is the (Poisson's) vector equation governing $\overline{A}$. In a system of rectangular coordinates it corresponds to the following three scalar equations

$$\begin{aligned}\left(\overline{\nabla}^2\overline{A}\right)_x &= \frac{\partial^2 A_x}{\partial x^2} + \frac{\partial^2 A_x}{\partial y^2} + \frac{\partial^2 A_x}{\partial z^2} = -\mu J_x \\ \left(\overline{\nabla}^2\overline{A}\right)_y &= -\mu J_y \\ \left(\overline{\nabla}^2\overline{A}\right)_z &= -\mu J_z\end{aligned} \tag{4.23}$$

In general, the gradient of an harmonic function η may be added to $\overline{A}$, having all the equations fulfilled. Suitable boundary conditions on Γ must be added in order to define the field in a unique way.

In particular, after (4.18) and (4.22), the potential inside a permanent magnet is given by $\overline{\nabla}^2\overline{A} = -\overline{\nabla} \times \overline{B}_0$.

(ii) In a two-dimensional domain, vectors $\overline{J}$ and so $\overline{A}$ have only one non-zero component; hence, vector potential can be treated as a scalar quantity.

Boundary conditions (4.8) and (4.10), in terms of $\overline{B} = \left(B_x, B_y\right)$ along the boundary Γ with normal unit vector $\overline{n} = \left(n_x, n_y\right)$ and tangential unit vector $\overline{t} = \left(t_x, t_y\right) = \left(n_y, -n_x\right)$, become, in terms of A,

$$\begin{aligned}\overline{n} \cdot \overline{B} &= n_x B_x + n_y B_y = n_x \frac{\partial A}{\partial y} - n_y \frac{\partial A}{\partial x} \\ &= -t_y \frac{\partial A}{\partial y} - t_x \frac{\partial A}{\partial x} = -\overline{t} \cdot \overline{\nabla} A = -\frac{\partial A}{\partial t} = 0\end{aligned} \tag{4.24}$$

i.e. $A = \text{const}$ along Γ and

$$\begin{aligned}\overline{n} \times \overline{B} &= \left(n_x B_y - n_y B_x\right)\overline{i}_z = \left(-n_x \frac{\partial A}{\partial x} - n_y \frac{\partial A}{\partial y}\right)\overline{i}_z \\ &= -\left(\overline{n} \cdot \overline{\nabla} A\right)\overline{i}_z = -\frac{\partial A}{\partial n}\overline{i}_z = 0\end{aligned} \tag{4.25}$$

i.e. $\frac{\partial A}{\partial n} = 0$ along Γ, respectively.

(iii) If $\overline{J} = 0$ in Ω and Ω is simply connected, then, along with $\overline{A}$, the field $\overline{H}$ can be described by a scalar function φ (total scalar potential, A) defined as

$$\overline{H} = -\overline{\nabla}\varphi \tag{4.26}$$

In fact, (4.3) is automatically satisfied, while from (4.2) we obtain

$$\overline{\nabla} \cdot \mu\overline{\nabla}\varphi = 0 \quad \text{in } \Omega \tag{4.27}$$

The latter is the Laplace's equation governing total scalar potential φ with suitable boundary conditions.

The condition of simply connected domain can be obtained by suitable cuts, if necessary. If this condition is not fulfilled, nevertheless φ can be still defined, apart from multiples of a constant.

(iv) In a three-dimensional domain, following the $\overline{T} - \Omega$ method, in regions free of impressed current ($J_0 = 0$) an electric vector potential $\overline{T}$ (A m^{-1}) can be defined as

$$\overline{\nabla} \times \overline{T} = \overline{J} \tag{4.28}$$

Comparing (4.28) and (4.3) it turns out that $\overline{H}$ and $\overline{T}$, which have the same curl, must differ by the gradient of a function Ω (dual scalar potential, A)

$$\overline{H} = \overline{T} - \overline{\nabla}\Omega \tag{4.29}$$

This way, the electric and magnetic vectors, $\overline{J}$ and $\overline{H}$, have been reformulated in terms of two potentials.

In order to define $\overline{T}$ uniquely, a gauge must be introduced.

The equations governing electric and magnetic field can be now expressed in terms of $\overline{T}$ and Ω. In fact, from (4.3) taking the curl of both members and taking into account (4.2) and (4.29), one has

$$\begin{aligned} \overline{\nabla} \times \left(\sigma^{-1}\overline{\nabla} \times \overline{T}\right) &= \overline{\nabla} \times \sigma^{-1}\overline{J}_0 - \frac{\partial}{\partial t}\mu\left(\overline{T} - \overline{\nabla}\Omega\right) \\ \overline{\nabla} \times \left(\sigma^{-1}\overline{\nabla} \times \overline{T}\right) &= \overline{\nabla} \times \sigma^{-1}\overline{J}_0 \end{aligned} \tag{4.30}$$

and from (4.2)

$$\overline{\nabla} \cdot \mu\left(\overline{T} - \overline{\nabla}\Omega\right) = 0 \tag{4.31}$$

In regions where $\sigma = 0$ one has $\overline{J} = 0$ and therefore, from (4.2), $\overline{\nabla} \times \overline{T} = 0$.

Moreover, imposing the gauge $\overline{\nabla} \cdot \overline{T} = \mu\sigma \frac{\partial \Omega}{\partial t}$ $\overline{\nabla} \cdot \overline{T} = 0$, from (4.30) and (4.31) one obtains two independent equations for T and Ω, namely

$$\overline{\nabla}^2 \overline{T} - \mu\sigma \frac{\partial \overline{T}}{\partial t} = -\overline{\nabla} \times \overline{J}_0$$
$$\overline{\nabla}^2 \overline{T} = -\overline{\nabla} \times \overline{J}_0 \tag{4.32}$$

and

$$\nabla^2 \Omega - \mu\sigma \frac{\partial \Omega}{\partial t} = 0$$
$$\nabla^2 \Omega = 0 \tag{4.33}$$

subject to appropriate boundary conditions. They are

$$\overline{n} \times \overline{T} = 0, \quad \Omega = 0 \tag{4.34}$$

or

$$\overline{n} \cdot \overline{T} = 0, \quad \frac{\partial \Omega}{\partial n} = 0 \tag{4.35}$$

if the boundary is normal to a flux line (i.e. $\overline{n} \times \overline{B} = 0$) or it is parallel to a flux line (i.e. $\overline{n} \cdot \overline{B} = 0$), respectively.

After determining $\overline{T}$, Ω is given by

$$\Omega(t) = \Omega_0 + (\mu\sigma)^{-1} \int_0^t \overline{\nabla} \cdot \overline{T}(t')dt' \tag{4.36}$$

with Ω_0 to be determined.

The following remark can be put forward.

In the two-dimensional case, the magnetic vector potential has only one non-zero component, and this makes the computational cost low. In contrast, if a formulation based on magnetic vector potential is used for a three-dimensional problem, all three vector components are unknown; therefore, the computational cost is high. The $\overline{T}$-Ω formulation is a good compromise: in fact, the use of vector potential $\overline{T}$ in current-carrying conductors makes it possible an accurate modeling of current distribution, while the use of scalar potential Ω elsewhere leads to economy in computation. Suitable conditions are needed at the boundary between conducting and non-conducting materials.

(v) When in (4.1) permeability μ depends on $|\overline{H}|$, one has $|\overline{B}| = \mu(|\overline{H}|)|\overline{H}|$ and for the solution of (4.22) one should resort to an iterative procedure. According e.g. to the Newton-Raphson method, the residual *r(A)* of the governing Eq. (4.22) is developed in Taylor's series, truncating the development at the first order

$$r(A_k) = r(A_{k-1}) + \left(\left. \frac{dr}{dA} \right|_{A=A_{k-1}} \right) (A_k - A_{k-1}) + o(A_k) \tag{4.37}$$

If an estimate of solution A_{k-1} at the (k − 1)-th iteration is available, the subsequent prediction A_k at the k-th iteration is given by (4.28) after imposing $r(A_k) = 0$. It turns out to be

$$A_k = A_{k-1} - \left[\left. \frac{dr}{dA} \right|_{A=A_{k-1}} \right]^{-1} r(A_{k-1}) \tag{4.38}$$

Then, μ and so $|\overline{H}|$ are updated by means of the new estimate of A, and the problem is solved again. The procedure stops when the error between two successive solutions is less than the prescribed threshold. It is necessary to know an initial prediction A_0 and the value of the derivative $\frac{dr}{dA}$ at each iteration.

4.3.1 *Field of a Line Current in a Three-Dimensional Domain: Differential Approach*

A current $I(A)$, concentrated at $r = 0$ and directed along the z axis in a system of cylindrical coordinates (r, φ, z), is considered (Fig. 4.2) [1].

The symmetry implies $\overline{H} = (0, H, 0)$ and from (4.3) the field equation is

$$\overline{\nabla} \times \overline{H} = \frac{1}{r} \frac{\partial rH}{\partial r} = \frac{\partial H}{\partial r} + \frac{1}{r} H = I\delta(r), \quad r > 0 \tag{4.39}$$

where H vanishes as r approaches infinity. The general solution is

$$H(r) = \frac{1}{r} \left(I \int_0^r \rho \delta(\rho) d\rho + k \right) \tag{4.40}$$

The Dirac's δ in a cylindrical geometry can be approximated by

$$\delta = \lim_{\alpha \to 0} \delta_\alpha, \quad \alpha > 0 \tag{4.41}$$

with $\delta_\alpha = \frac{1}{\pi \alpha^2}$, $r \le \alpha$ and $\delta_\alpha = 0$ elsewhere. Consequently, the field H can be approximated as

$$H = \lim_{\alpha \to 0} H_\alpha \tag{4.42}$$

For $r \le \alpha$ it turns out to be

$$H_\alpha = \frac{1}{r}\left(I \int_0^r \rho\delta_\alpha d\rho + k_\alpha \right) = \frac{1}{r}\left(\frac{I}{\pi\alpha^2}\frac{r^2}{2} + k_\alpha \right)$$
$$= \frac{Ir}{2\pi\alpha^2} + \frac{k_\alpha}{r} \tag{4.43}$$

Since δ_α is a regular function near the origin, also H_n will be regular near zero; therefore $k_\alpha = 0$.

For $r \geq \alpha$ it turns out to be

$$H_\alpha = \frac{1}{r}\left(I \int_0^\alpha \rho\delta_\alpha d\rho + k_\alpha \right) = \frac{1}{r}\left(\frac{I}{\pi\alpha^2} \int_0^\alpha \rho d\rho \right)$$
$$= \frac{1}{r}\left(\frac{I}{\pi\alpha^2}\frac{\alpha^2}{2} \right) = \frac{I}{2\pi r}, \quad r > 0 \tag{4.44}$$

The Biot-Savart's law follows

$$H(r) = \lim_{\alpha \to 0} H_\alpha(r) = \frac{I}{2\pi r}, \quad r > 0 \tag{4.45}$$

Alternatively, the Stokes's theorem can be applied to (4.3), giving $\oint_l \overline{H} \cdot \overline{t} d\ell = I$, if ℓ is a closed line linking the conductor once. Considering the field geometry, ℓ can be taken as a circular line centred at $r = 0$; therefore, (4.41) follows.

From (4.41) and (4.19) the vector potential is

$$\overline{A} = \frac{I}{2\pi\nu} \ln r \overline{i}_z, \quad r > 0 \tag{4.46}$$

4.4 Magnetostatic Energy

Given a magnetostatic field characterized by strength $\overline{H}$ and flux density $\overline{B}$ in a linear medium, the specific energy (J m^{-3}) of the field is defined as $\frac{1}{2}\overline{H} \cdot \overline{B}$; if the medium is isotropic, the energy W(J) stored in an unbounded region Ω is given by

$$W = \frac{1}{2} \int_\Omega HB d\Omega \tag{4.47}$$

If the constitutive relationship of the magnetic material is non-linear, the specific energy is $\int_0^B H\, dB'$ and the total energy is

$$W = \int_{\Omega} \left(\int_{0}^{B} H \, dB' \right) d\Omega \tag{4.48}$$

In some cases it is convenient to introduce the specific co-energy $\int_0^H B \, dH'$ and the total co-energy is

$$W' = \int_{\Omega} \left(\int_{0}^{H} B \, dH' \right) d\Omega \tag{4.49}$$

In the case of linear medium $W = W'$ holds.

In the linear case, taking into account the following identity (see A.13)

$$\overline{H} \cdot \overline{B} = \overline{H} \cdot \left(\overline{\nabla} \times \overline{A}\right) = \overline{A} \cdot \left(\overline{\nabla} \times \overline{H}\right) - \overline{\nabla} \cdot \left(\overline{H} \times \overline{A}\right) = \overline{A} \cdot \overline{J} - \overline{\nabla} \cdot \left(\overline{H} \times \overline{A}\right) \tag{4.50}$$

and (4.3), the total energy stored in a region Ω of boundary Γ is

$$W = \frac{1}{2} \int_{\Omega} \overline{H} \cdot \overline{B} \, d\Omega = \frac{1}{2} \int_{\Omega} \overline{A} \cdot \overline{J} \, d\Omega - \frac{1}{2} \int_{\Gamma} \left(\overline{H} \times \overline{A}\right) \cdot \overline{n} \, d\Gamma \tag{4.51}$$

The equation above reduces to $W = \frac{1}{2} \int_{\Omega} \overline{A} \cdot \overline{J} \, d\Omega$ if either $\overline{A} \times \overline{n} = 0$ or $\overline{H} \times \overline{n} = 0$ along Γ [3].

4.5 Forces and Torques in the Magnetostatic Field

4.5.1 Principle of Virtual Work

Given a structure in the field region, on which force $\overline{F}$ is to be computed, a virtual linear displacement ds in the direction of $\overline{F}$, supposing that the magnetic system is supplied by a constant current I creating a linkage flux Φ, the sum of mechanical work Fds and variation of magnetic energy dW is equal to the input energy $Id\Phi$ so that the following balance equation

$$\begin{aligned} F \, ds + dW &= I d\Phi \\ F \, ds &= d(I\Phi - W) \\ F &= \frac{d}{ds}(I\Phi - W) \end{aligned} \tag{4.52}$$

In the case of an angular displacement $d\vartheta$, the torque M with respect to the rotation axis is

$$M = \frac{d}{d\vartheta}(I\Phi - W) \tag{4.53}$$

The quantity $I\Phi$-W, denoted by W', is the complementary energy or co-energy of the system.

On the other hand, if the magnetic system is isolated, mechanical work Fds and variation of magnetic energy dW take place so that

$$F\,ds + dW = 0 \tag{4.54}$$

Therefore, the force can be evaluated as

$$F = -\frac{dW}{ds} \tag{4.55}$$

while the torque is

$$M = -\frac{dW}{d\vartheta} \tag{4.56}$$

If the system is linear, W' and W coincide.

4.5.2 Lorentz's Method

It is based on the definition of flux density; in the empty space, the force $\overline{F}$ exerted on current I carried by a linear conductor of length ℓ is $\overline{F} = I\overline{\ell} \times \overline{B}$ where $\overline{B}$ is the external field, i.e. the flux density in the absence of current. In general, the force $\overline{F}$ exerted on current distributed with density $\overline{J}$ in the region Ω is

$$\overline{F} = \int_{\Omega} \overline{J} \times \overline{B}\, d\Omega \tag{4.57}$$

Direction of force is orthogonal to the plane defined by flux density and current density vectors.

4.5.3 Method of Maxwell's Stress Tensor

Defined a closed surface Γ enclosing the structure, then force $\overline{F}$ is evaluated as

$$\overline{F} = \int_{\Omega} \nabla \cdot \overline{\overline{T}}\, d\Omega = \int_{\Gamma} \overline{\overline{T}} \cdot \bar{n}\, d\Gamma \tag{4.58}$$

where $\bar{n}$ is the outward normal unit vector.

The Maxwell's magnetic stress tensors $\overline{\overline{T}}$, assuming a system of rectangular coordinates, in a three-dimensional domain can be represented in matrix form as

$$\overline{\overline{T}} = \begin{bmatrix} \frac{1}{2}(H_xB_x - H_yB_y - H_zB_z) & H_xB_y & H_xB_z \\ H_yB_x & \frac{1}{2}(H_yB_y - H_xB_x - H_zB_z) & H_yB_z \\ H_zB_x & H_zB_y & \frac{1}{2}(H_zB_z - H_xB_x - H_yB_y) \end{bmatrix} \tag{4.59}$$

In order the tensor be uniquely defined, surface Γ should not be coincident with the interface between materials having different permeability [2].

4.5.4 Link Between Lorentz's and Maxwell's Approach

There is a link between Lorentz's and Maxwell's approach to force calculation. In fact, using (4.1), (4.3) and (4.57), the force density $\overline{f}\,(\mathrm{Nm}^{-3})$ is

$$\overline{f} = \overline{J} \times \overline{B} = \left(\overline{\nabla} \times \nu\overline{B}\right) \times \overline{B} \tag{4.60}$$

In particular, the x-directed component is

$$f_x = \nu B_z \frac{\partial B_x}{\partial z} - \nu B_z \frac{\partial B_z}{\partial x} - \nu B_y \frac{\partial B_y}{\partial x} + \nu B_y \frac{\partial B_x}{\partial y} \tag{4.61}$$

After adding and subtracting the term $\frac{\nu}{2}\frac{\partial B_x^2}{\partial x}$ it follows

$$\begin{aligned} f_x = {} & \frac{\nu}{2}\frac{\partial B_x^2}{\partial x} + \nu B_z \frac{\partial B_x}{\partial z} + \nu B_y \frac{\partial B_x}{\partial y} + \\ & -\frac{\nu}{2}\frac{\partial}{\partial x}\left(B_x^2 + B_y^2 + B_z^2\right) \end{aligned} \tag{4.62}$$

It turns out to be

$$f_x = \frac{\nu}{2}\frac{\partial B_x^2}{\partial x} + \nu\frac{\partial(B_x B_z)}{\partial z} - \nu B_x\frac{\partial B_z}{\partial z} + \nu\frac{\partial(B_x B_y)}{\partial y} + \\ - \nu B_x\frac{\partial B_y}{\partial y} - \nu B_x\frac{\partial B_x}{\partial x} - \frac{\nu}{2}\frac{\partial}{\partial x}\left(B_y^2 + B_z^2\right) \tag{4.63}$$

$$f_x = \nu\left[\frac{\partial}{\partial x}\left(B_x^2 - \frac{1}{2}\left|\overline{B}\right|^2\right) + \frac{\partial(B_x B_y)}{\partial y} + \frac{\partial(B_x B_z)}{\partial z} - B_x\overline{\nabla}\cdot\overline{B}\right] \tag{4.64}$$

Due to (4.2) the last term of (4.64) is zero; then, if vector

$$\overline{v}_1 = \nu\left(B_x^2 - \frac{1}{2}\left|\overline{B}\right|^2, B_x B_y, B_x B_z\right) \\ = \left(\frac{1}{2}\left(H_x B_x - H_y B_y - H_z B_z\right), H_x B_y, H_x B_z\right) \tag{4.65}$$

is defined, f_x can be viewed as its divergence, apart from a constant k which can be set to zero, namely

$$f_x = \overline{\nabla}\cdot\overline{v}_1 \tag{4.66}$$

A similar result holds for force density components f_y and f_z; it follows

$$\overline{v}_2 = \left(H_y B_x, \frac{1}{2}\left(H_y B_y - H_x B_x - H_z B_z\right), H_y B_z\right) \tag{4.67}$$

such that

$$f_y = \overline{\nabla}\cdot\overline{v}_2 \tag{4.68}$$

and

$$\overline{v}_3 = \left(H_z B_x, H_z B_y, \frac{1}{2}\left(H_z B_z - H_x B_x - H_y B_y\right)\right) \tag{4.69}$$

such that

$$f_z = \overline{\nabla}\cdot\overline{v}_3 \tag{4.70}$$

respectively. Therefore, according to (4.58), the force $\overline{F}(N)$ can be computed as the flux, leaving surface Γ, of tensor $\overline{\overline{T}}$ represented by matrix (4.59), in which the row entries are the components of vectors $\overline{v}_k$, $k = 1, 3$.

Correspondingly, the torque is given by $\overline{M} = \int_\Gamma \overline{r}_{PO} \times \overline{\overline{T}} \cdot \overline{n}\, d\Gamma$ where $\overline{r}_{PO}$ is the position vector of point P on Γ with respect to the rotation axis in O.

It can be remarked that a solenoidal vector $\overline{w}$ may be added to (4.65), (4.67) and (4.69) leaving force density components (4.66), (4.68) and (4.70) fulfilled. This means that stress tensor (4.59) is not uniquely defined.

As far as a comparison of methods is concerned, the following remark can be put forward. In order Lorentz's method to apply, a current density must be defined in Ω; in contrast, virtual work principle (VWP) and Maxwell's stress tensor method (MST) are more general. VWP is computationally more expensive, because the derivative of energy or co-energy is approximated by means of a finite difference, involving two displaced positions of the structure. Therefore, two field analyses are necessary to compute force or torque at a given position. MST require only one field analysis.

4.6 Worked Example

4.6.1 *Force on an Electromagnet*

Let an electromagnet with a movable plunger be considered (Fig. 4.3), [1].

The iron core is supposed to have infinite permeability. The air gaps in the x direction are supposed to be much smaller than the air gap t in the y direction.

The circulation of the magnetic field H, along a line linking the excitation current NI and crossing the air gap t in the normal direction, reduces to

$$NI = Ht \tag{4.71}$$

Therefore at the air gap

$$H = \frac{NI}{t} \tag{4.72}$$

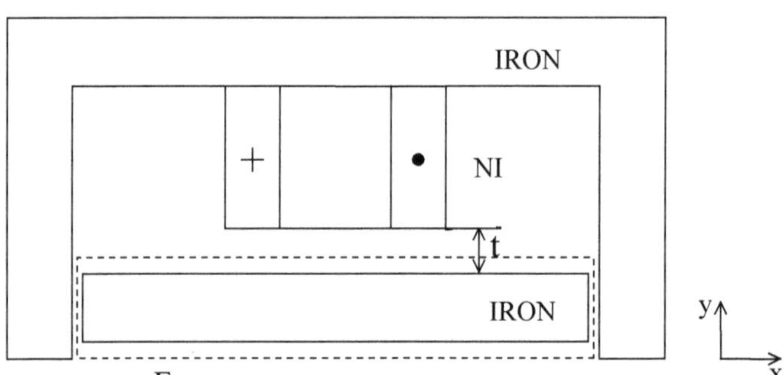

Fig. 4.3 Model of the electromagnet

while in the iron part $H = 0$. Following (4.49), the co-energy stored in the air gap is given by

$$W' = \frac{1}{2}\mu_0 H^2 St = \frac{\mu_0 (NI)^2 S}{2t} \tag{4.73}$$

where S is the cross-section of the central limb and μ_0 is the air permeability.

If NI is constant, according to (4.52), the force acting on the movable part is

$$F_t = \frac{\partial W'}{\partial t} = -\frac{\mu_0 S}{2}\left(\frac{NI}{t}\right)^2 \tag{4.74}$$

The force is negative, i.e. opposite to the direction of increasing t; therefore, it is attractive, regardless of the sign of I.

In order to apply the method of Maxwell's stress tensor, an integration surface Γ enclosing the movable part is considered having $\overline{n}$ as its outward normal unit vector.

Taking into account the field distribution, it follows

$$\overline{\overline{T}} = \begin{bmatrix} -\frac{1}{2}H_y B_y & 0 \\ 0 & \frac{1}{2}H_y B_y \end{bmatrix} \tag{4.75}$$

$$\overline{F} = \int_\Gamma \overline{\overline{T}} \cdot \overline{n}\, d\Gamma = \left(0, \frac{1}{2}H_y B_y S\right) \tag{4.76}$$

Therefore it turns out to be

$$F_y = \frac{1}{2}\mu_0 H_y^2 S = \frac{1}{2}\mu_0 S\left(\frac{NI}{t}\right)^2 \tag{4.77}$$

The force is attractive, because variables t and y are oriented in opposite directions.

References

1. Di Barba P, Savini A, Wiak S (2008) Field models in electricity and magnetism. Springer, Berlin, Germany
2. Hammond P (1971) Applied electromagnetism. Pergamon
3. Silvester PP (1968) Modern electromagnetic fields. Prentice-Hall
4. Simonyi K (1963) Foundations of electrical engineering: fields, networks, waves. Pergamon

Chapter 5
Steady Conduction Field and Boundary-Value Problems

5.1 Constitutive Law of Conducting Material

In a domain Ω, having boundary Γ, filled in by a conducting material, when a voltage—constant with time—is set up across Γ, a steady conduction field originates; it is defined by field strength $\overline{E}$ (V m^{-1}) and current density $\overline{J}$ (A m^{-2}). These two vectors are linked by the constitutive relation which, if the medium is linear and isotropic, is

$$\overline{J} = \sigma \overline{E} \tag{5.1}$$

The parameter σ qualifying the material is called conductivity (Ω^{-1} m^{-1}) [4].

5.2 Maxwell's Equations of Conduction Field

The field is governed by the following Maxwell's equations in Ω

$$\overline{\nabla} \times \overline{E} = 0 \tag{5.2}$$

$$\overline{\nabla} \cdot \overline{J} = 0 \tag{5.3}$$

and along Γ

$$\overline{n} \cdot \overline{J} = J_0, \quad \overline{n} \times \overline{J} = 0 \tag{5.4}$$

if Γ is a perfectly conducting boundary and J_0 is the current density impressed along it, or

$$\overline{n} \cdot \overline{J} = 0 \tag{5.5}$$

P. Di Barba and S. Wiak, *MEMS: Field Models and Optimal Design*, Lecture Notes in Electrical Engineering 573, https://doi.org/10.1007/978-3-030-21496-8_5

if Γ is a perfectly insulating boundary.

In terms of $\overline{E}$ the above equations for a homogeneous medium become in Ω

$$\overline{\nabla} \times \overline{E} = 0 \tag{5.6}$$

$$\overline{\nabla} \cdot \overline{E} = 0 \tag{5.7}$$

and along Γ

$$\overline{n} \cdot \overline{E} = E_0, \quad \overline{n} \times \overline{E} = 0 \tag{5.8}$$

if E_0 is the conduction field impressed along the boundary, or

$$\overline{n} \cdot \overline{E} = 0 \tag{5.9}$$

if Γ is a perfectly insulating boundary.

In the case of a non-homogeneous medium, the following remark can be put forward.

After (5.1) and (5.2), considering vector identity (A.16), it turns out to be

$$\overline{\nabla} \times \sigma^{-1}\overline{J} = \sigma^{-1}\overline{\nabla} \times \overline{J} + \overline{\nabla}\sigma^{-1} \times \overline{J} = 0 \tag{5.10}$$

If $\overline{\nabla}\sigma^{-1} = 0$ (homogeneous medium), then $\overline{J}$ is always irrotational. More generally, for $\overline{J}$ to be irrotational, $\overline{\nabla}\sigma^{-1}$ and $\overline{J}$ should be parallel vectors; this means that lines separating layers of different σ are orthogonal to field lines of $\overline{J}$. If this condition applies, $\overline{J}$ is both irrotational and solenoidal.

Conversely, after (5.1) and (A.14), it follows

$$\overline{\nabla} \cdot \overline{E} = \overline{\nabla} \cdot \sigma^{-1}\overline{J} = \sigma^{-1}\overline{\nabla} \cdot \overline{J} + \overline{\nabla}\sigma^{-1} \cdot \overline{J} = 0 \tag{5.11}$$

Due to (5.3), it follows

$$\overline{\nabla} \cdot \overline{E} = \overline{\nabla}\sigma^{-1} \cdot \overline{J} = 0 \tag{5.12}$$

i.e. field $\overline{E}$ is solenoidal if $\overline{\nabla}\sigma^{-1}$ and $\overline{J}$ are orthogonal vectors; this means that lines separating layers of different σ are parallel to field lines of $\overline{J}$. If this condition applies, $\overline{E}$ is both irrotational and solenoidal [4].

5.3 From Field to Potentials

If Ω is simply connected, starting from (5.2) the field can be defined by a scalar function U [potential (V)] as

$$\overline{E} = -\overline{\nabla} U \tag{5.13}$$

Starting from (5.3), substitution of (5.13) in it, taking into account (5.1), gives

$$\overline{\nabla} \cdot \left(\sigma \overline{\nabla} U \right) = 0 \tag{5.14}$$

In rectangular coordinates, (5.14) becomes

$$\frac{\partial}{\partial x}\left(\sigma \frac{\partial U}{\partial x} \right) + \frac{\partial}{\partial y}\left(\sigma \frac{\partial U}{\partial y} \right) + \frac{\partial}{\partial z}\left(\sigma \frac{\partial U}{\partial z} \right) = 0 \tag{5.15}$$

In the case of a homogeneous domain (5.14) becomes

$$\nabla^2 U = 0 \tag{5.16}$$

which is Laplace's equation governing potential U.

Any constant may be added to U, keeping all the equations valid.

In particular, in a two-dimensional domain, using rectangular coordinates, the field strength $\overline{E} = \left(E_x, E_y \right)$ defined by (5.13) has components

$$E_x = -\frac{\partial U}{\partial x}, \quad E_y = -\frac{\partial U}{\partial y} \tag{5.17}$$

Conversely, starting from (5.3), according to (A.8), it is possible to introduce a vector function $\overline{A}$ [flux (A m^{-1})] defined as

$$\overline{J} = \overline{\nabla} \times \overline{A} \tag{5.18}$$

In a two-dimensional domain, considering the definition of curl as well of that of current density $\overline{J} = (J_x, J_y)$, $\overline{A}$ turns out to be a single-component vector orthogonal to the domain, namely $\overline{A} = (0, 0, A)$. Therefore, the components of $\overline{J}$ turn out to be

$$J_x = \frac{\partial A}{\partial y}, \quad J_y = -\frac{\partial A}{\partial x} \tag{5.19}$$

Substituting (5.18) into (5.2), after multiplication by σ and taking into account (5.1), one obtains

$$\overline{\nabla} \times \sigma^{-1} \overline{\nabla} \times \overline{A} = 0 \tag{5.20}$$

In a two-dimensional domain this, in turn, becomes

$$\frac{\partial}{\partial x}\left(\sigma^{-1}\frac{\partial A}{\partial x}\right)+\frac{\partial}{\partial y}\left(\sigma^{-1}\frac{\partial A}{\partial y}\right)=0 \tag{5.21}$$

In the case of homogeneous domain (5.21) reduces to

$$\nabla^2 A = 0 \tag{5.22}$$

which is Laplace's equation governing flux A.

The gradient of an harmonic function η, having all the equations fulfilled as well, may be added to $\overline{A}$; in fact, by imposing

$$\overline{\nabla}\cdot\overline{A} = 0 \text{ (gauge condition)} \tag{5.23}$$

$\overline{A}$ is unambiguously defined. In a two-dimensional domain the latter condition is automatically fulfilled.

Comparing (5.17) and (5.19) it turns out to be

$$\frac{\partial A}{\partial y} = -\sigma\frac{\partial U}{\partial x};\quad \frac{\partial A}{\partial x} = \sigma\frac{\partial U}{\partial y} \tag{5.24}$$

The latter represent relationships of orthogonality between contour lines of the two potentials.

Boundary conditions (5.8) and (5.9) become in terms of potential $U = \text{const.}$ and $\frac{\partial U}{\partial n} = 0$ respectively; in turn, (5.8) and (5.9) become in terms of flux $\frac{\partial A}{\partial n} = 0$ and $A = \text{const.}$, respectively [3].

5.4 Power Loss

Given a conduction field characterized by strength $\overline{E}$ and current density $\overline{J}$, the specific power (W m^{-3}) transferred from the field to movables charges is defined as $\overline{E}\cdot\overline{J} = \sigma E^2$ and, therefore, the power loss (W) in domain Ω is given by:

$$P = \int_{\Omega} \sigma E^2 d\Omega \tag{5.25}$$

Equation (5.25) models the Joule's law [2].

5.4.1 Field of a Cylindrical Conductor

Let a cylindrical homogenous conductor of radii R_1 and R_2, carrying current I per unit length in the radial direction, be considered (Fig. 5.1) [1].

From (5.16), according to (A.18), it follows

$$\nabla^2 U = \frac{d^2U}{dr^2} + r^{-1}\frac{dU}{dr} = 0, \quad r > 0 \tag{5.27}$$

and therefore

$$U = k \ln r + h \tag{5.28}$$

The application of boundary conditions

$$r = R_1, \quad U = 0; \quad r = R_2, \quad U = V \tag{5.29}$$

gives

$$U = V\frac{\ln r - \ln R_1}{\ln R_2 - \ln R_1}, \quad R_1 \le r \le R_2 \tag{5.30}$$

and

$$\overline{E} = -\frac{V}{r(\ln R_2 - \ln R_1)}\overline{i}_r, \quad R_1 \le r \le R_2 \tag{5.31}$$

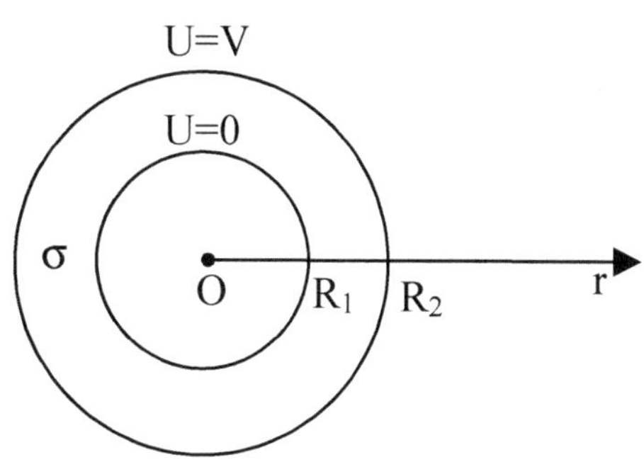

Fig. 5.1 Cross-section of a cylindrical conductor

In the formulation in terms of flux A, after making a radial cut at $\vartheta = 0$ in order to have a simply connected domain, the boundary conditions are

$$\vartheta = 0, \quad A = 0; \quad \vartheta = 2\pi, \quad A = I \tag{5.32}$$

$$r = R_1, \quad \frac{\partial A}{\partial n} = 0; \quad r = R_2, \quad \frac{\partial A}{\partial n} = 0 \tag{5.33}$$

From (5.22), it follows

$$r^{-2}\frac{d^2 A}{d\vartheta^2} = 0, \quad r > 0 \tag{5.34}$$

and therefore

$$\mathrm{A = k\vartheta + h} \tag{5.35}$$

By the application of boundary conditions, it turns out to be

$$A = \frac{I}{2\pi}\vartheta, \quad 0 \le \vartheta \le 2\pi \tag{5.36}$$

and

$$\overline{J} = -\frac{I}{2\pi r}\overline{i}_r, \quad R_1 \le r \le R_2 \tag{5.37}$$

After (5.1) it turns out to be

$$\frac{I}{2\pi r} = \frac{\sigma V}{r(\ln\, R_2 - \ln\, R_1)}, \quad R_1 \le r \le R_2 \tag{5.38}$$

Let the equivalent resistance R_{eq} of the conductor be defined as the voltage-to-current ratio; then, from (5.38) it follows

$$R_{eq} = \frac{V}{I} = \frac{\ln\, R_2 - \ln\, R_1}{2\pi\sigma} = \frac{1}{2\pi\sigma}\ln\left(\frac{R_2}{R_1}\right) \tag{5.39}$$

Finally, after (5.25), the power loss is given by

$$P = \int_{R_1}^{R_2} EJ2\pi r dr = \frac{VI}{\ln\, R_2 - \ln\, R_1}\int_{R_1}^{R_2}\frac{dr}{r} = VI \tag{5.40}$$

References

1. Di Barba P, Savini A, Wiak S (2008) Field models in electricity and magnetism. Springer, Berlin, Germany
2. Hammond P (1971) Applied electromagnetism. Pergamon
3. Silvester PP (1968) Modern electromagnetic fields. Prentice-Hall
4. Simonyi K (1963) Foundations of electrical engineering: fields, networks, waves. Pergamon

Chapter 6
From Fields to Circuits

6.1 Maxwell's Equations of Electromagnetic Field

In general, the twofold presence of a movable charge distributed with density ρ ($\mathrm{C\,m^{-3}}$) and an impressed current density $\overline{J}_0$ ($\mathrm{A\,m^{-2}}$) variable with time gives origin to the electromagnetic field described by the following time-dependent vectors:

$\overline{D}$ electric flux density ($\mathrm{C\ m^{-2}}$)
$\overline{E}$ electric field strength ($\mathrm{V\ m^{-1}}$)
$\overline{B}$ magnetic flux density (T)
$\overline{H}$ magnetic field strength ($\mathrm{A\ m^{-1}}$)
$\overline{J}$ current density ($\mathrm{A\ m^{-2}}$)

As far as the origin of current density is concerned, the following remark can be put forward. In a solid or liquid medium the *conduction current* density is a function of $\overline{E}$ field

$$\overline{J} = \overline{J}(\overline{E}) \tag{6.1}$$

For a linear material the above function becomes

$$\overline{J} = \sigma \overline{E} \tag{6.2}$$

Another kind of current is originated by the movement of free ions and electrons (e.g. in gases or vacuum). This *convection current* density is expressed by the formula

$$\overline{J} = \rho_+ u_+ + \rho_- u_- \tag{6.3}$$

where ρ_+ and ρ_- are positive and negative charge densities, respectively, while u_+ and u_- are the relevant velocities of positive and negative free charges.

P. Di Barba and S. Wiak, *MEMS: Field Models and Optimal Design*, Lecture Notes in Electrical Engineering 573, https://doi.org/10.1007/978-3-030-21496-8_6

Finally, the *displacement current* density is defined as

$$\overline{J} = \frac{\partial \overline{D}}{\partial t} \tag{6.4}$$

Considering the principle of charge conservation, in any point of the domain the following equation holds (charge continuity equation)

$$\overline{\nabla} \cdot \overline{J} + \frac{\partial \rho}{\partial t} = 0 \tag{6.5}$$

The coupled electric and magnetic fields influence a charge $q(C)$ by exerting a mechanical force $\overline{F}$ (N) on it (Lorentz's equation)

$$\overline{F} = q\left(\overline{E} + \bar{u} \times \overline{B}\right) \tag{6.6}$$

where $\overline{u}$ is the velocity of the charge with respect to the magnetic field. In particular, the term $q\overline{E}$ modifies the value of velocity, while the term $q\overline{u} \times \bar{B}$ modifies also the direction of velocity. In (6.6) $\overline{E}$ is the field in the absence of charge q: the latter can be considered as a probe testing the presence of $\overline{E}$ and $\overline{B}$ fields [1].

In a domain Ω with boundary Γ filled in by a linear medium characterized by permittivity ε, permeability μ and conductivity σ, the time-varying electromagnetic field is described by the following equations:

Faraday's equation

$$\overline{\nabla} \times \overline{E} = -\frac{\partial \overline{B}}{\partial t} \tag{6.7}$$

Gauss' electric equation

$$\overline{\nabla} \cdot \overline{D} = \rho \tag{6.8}$$

Ampère's equation

$$\overline{\nabla} \times \overline{H} = \overline{J} + \frac{\partial \overline{D}}{\partial t} \tag{6.9}$$

Gauss' magnetic equation

$$\overline{\nabla} \cdot \overline{B} = 0 \tag{6.10}$$

In a three-dimensional domain, the above equations represent a set of eight scalar equations to which constitutive relations (3.1), (4.1), (5.1) must be added.

In total, fifteen scalar unknowns (i.e. field components) have to be determined, subject to suitable boundary conditions.

The system of eight plus nine equations can be solved since there are two relations among the unknowns which are automatically satisfied. One is (6.5) and the other comes from (6.7) to (6.8). In fact, taking the divergence of (6.9) and the time derivative of (6.8), continuity equation (6.5) follows. Similarly, taking the divergence of (6.7) and the time derivative of (6.10), one obtains an identity.

It should be remarked that in (6.9), in general, the current density reads

$$\overline{J} = \bar{J}_0 + \sigma \overline{E} + \mu \sigma \overline{u} \times \overline{H} \tag{6.11}$$

where $\overline{J}_0$ is the term impressed by an external source, while the last term of the right-hand side takes into account the current density due to motional effect, if any.

In steady conditions all vectors are independent of time. Therefore, the two equations governing the electric field, namely (6.7) and (6.8), are decoupled with respect to the two equations governing the magnetic field, namely (6.9) and (6.10) (see Chaps. 3 and 4, respectively).

6.2 Poynting's Theorem

Let Maxwell's Eqs. (6.7) and (6.9) be considered. From vector identity (A.13) it follows

$$\nabla \cdot (\overline{E} \times \overline{H}) = \overline{H} \cdot (\overline{\nabla} \times \overline{E}) - \overline{E} \cdot (\overline{\nabla} \times \overline{H}) = -\overline{H} \cdot \frac{\partial \overline{B}}{\partial t} - \overline{E} \cdot \frac{\partial \overline{D}}{\partial t} - \overline{E} \cdot \overline{J} \tag{6.12}$$

Referring to the specific energy in the electric and magnetic case, and under the assumption of linear constitutive relationship, it follows

$$\begin{aligned}\frac{1}{2}\frac{\partial}{\partial t}(\overline{H} \cdot \overline{B} + \overline{E} \cdot \overline{D}) &= \frac{1}{2}\left(\overline{H} \cdot \frac{\partial \overline{B}}{\partial t} + \overline{B} \cdot \frac{\partial \overline{H}}{\partial t}\right) \\ &+ \frac{1}{2}\left(\overline{E} \cdot \frac{\partial \overline{D}}{\partial t} + \overline{D} \cdot \frac{\partial \overline{E}}{\partial t}\right) = \overline{H} \cdot \frac{\partial \overline{B}}{\partial t} + \overline{E} \cdot \frac{\partial \overline{D}}{\partial t}\end{aligned} \tag{6.13}$$

Integrating (6.12) over Ω and using Gauss's theorem (A.10), it turns out to be

$$\begin{aligned}\int_{\Gamma=\partial\Omega} (\overline{E} \times \overline{H}) \cdot \overline{n}\, d\Gamma &= -\frac{\partial}{\partial t}\int_{\Omega}\left(\frac{\overline{H} \cdot \overline{B}}{2} + \frac{\overline{E} \cdot \overline{D}}{2}\right) d\Omega - \int_{\Omega} \overline{E} \cdot \overline{J}\, d\Omega \\ &= -\frac{dW_m}{dt} - \frac{dW_e}{dt} - p_d\end{aligned} \tag{6.14}$$

Vector

$$\overline{S} = \overline{E} \times \overline{H} \tag{6.15}$$

is called Poynting's vector ($W\,m^{-2}$). According to (6.14), the flux of $\overline{S}$ leaving a closed surface Γ is equal to the decrease of both the power associated to the electromagnetic field and the power transferred to the current inside domain Ω bounded by Γ. This is the statement of Poynting's theorem.

If impressed current distributed with density $\overline{J}_0$ is present within volume Ω then (6.14) becomes

$$\int_{\Gamma} \left(\overline{E} \times \overline{H}\right) \cdot \overline{n}\, d\Gamma = -\frac{\partial}{\partial t} \int_{\Omega} \left(\frac{\overline{H} \cdot \overline{B}}{2} + \frac{\overline{E} \cdot \overline{D}}{2}\right) d\Omega$$
$$- \int_{\Omega} \overline{E} \cdot \overline{J}\, d\Omega + \int_{\Omega} \overline{E} \cdot \overline{J}_0\, d\Omega \tag{6.16}$$

where $p_g = -\int_{\Omega} \overline{E} \cdot \overline{J}_0\, d\Omega$ is the power delivered by impressed current density.

6.3 From Distributed to Lumped-Parameter Models

Maxwell's equations (6.7)–(6.10) can hardly be solved in an accurate way in most practical cases of real-life engineering, due to complicated geometries of devices and non-linearity of materials. However, if an electromagnetic device is characterized by slowly-varying fields, it can be modeled as a multi-terminal element, and this makes it possible to consider lumped-parameter models, i.e. circuits, which are fairly less costly to solve. On the other hand, however, circuits are zero-dimensional models and, therefore, the space dependence of electric and magnetic quantities is missing.

In fact, it is possible to move from field-based models—i.e. those governed by Maxwell's equations—to lumped-parameter models—i.e. those governed by Kirchhoff's equations—by exploiting Poynting's theorem. In particular, two-terminal elements, which are the basic components of any circuit, can be defined and classified according to (6.14).

In general, an electromagnetic device can be modelled as a lumped-parameter multi-terminal element if it can be embedded within a volume Ω bounded by a closed surface Γ, along which the normal components of time derivatives of fields can be neglected, namely

$$\frac{\partial \overline{D}}{\partial t} \cdot \overline{n} = 0, \quad \frac{\partial \overline{B}}{\partial t} \cdot \overline{n} = 0 \tag{6.17}$$

along Γ.

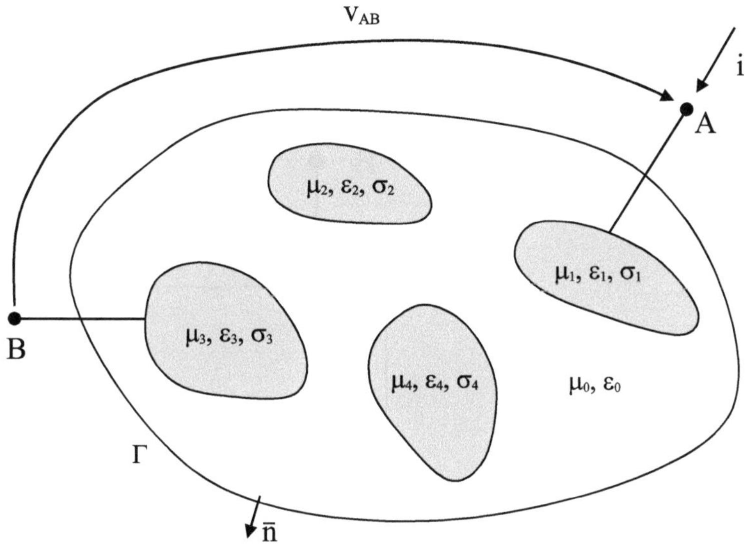

Fig. 6.1 From distributed to lumped parameter model: electromagnetic device embedded in a closed surface along which (6.17) holds

The simplest lumped-parameter model of an electromagnetic device is the two-terminal element (Fig. 6.1).

By means of the Poynting's theorem, the power entering a two-terminal element will be computed in the next section.

6.4 Power of a Two-Terminal Element

Let an element exhibiting terminals A and B be considered (Fig. 6.2).

Power $p(t)$ entering boundary surface S_c at time t is given by the flux of Poynting's vector (6.14)

$$p(t) = \int_{S_c} \left[\overline{E}(t) \times \overline{H}(t) \right] \cdot \overline{n} \, dS \tag{6.18}$$

Since, by assumption, Ω is a simply-connected domain and (6.17) holds, the electric voltage across two points on surface S_c is independent of any integration line joining them and belonging to S_c; therefore, a scalar function ψ (electric scalar potential) can be defined, such that the voltage across terminals A and B is given by

$$v_{AB} = \psi_A - \psi_B \tag{6.19}$$

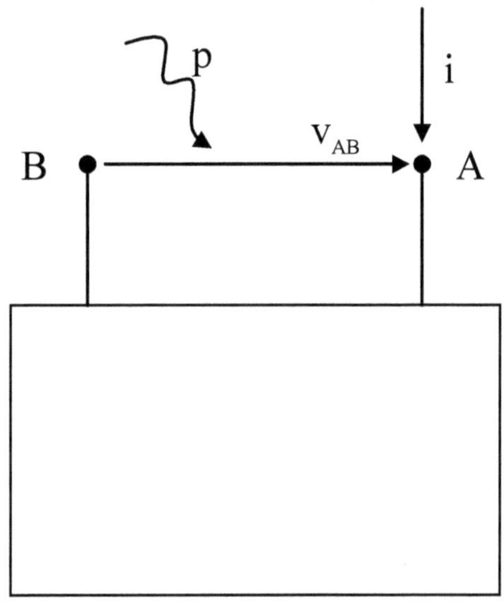

Fig. 6.2 Two-terminal element

Due to (6.7) and (6.17), the electric field is

$$\overline{E}(t) = -\overline{\nabla}\psi(t) \tag{6.20}$$

Then, the power is

$$p(t) = -\int_{S_c} \left[\overline{\nabla}\psi(t) \times \overline{H}(t)\right] \cdot \overline{n}\, dS \tag{6.21}$$

From vector identity (A.16) with $\phi = \psi$ and $\overline{V} = \overline{H}$ it follows

$$p(t) = -\int_{S_c} \left(\overline{\nabla} \times \psi\overline{H}\right) \cdot \overline{n}\, dS + \int_{S_c} \psi\overline{J} \cdot \overline{n}\, dS \tag{6.22}$$

Due to theorem (A.10) and identity (A.8) it turns out to be

$$p(t) = -\int_{\Omega} \nabla \cdot \left(\overline{\nabla} \times \psi\overline{H}\right) d\Omega + \int_{S_c=\partial\Omega} \psi\overline{J} \cdot \overline{n}\, dS = \int_{S_c=\partial\Omega} \psi\overline{J} \cdot \overline{n}\, dS \tag{6.23}$$

The only non-zero contributions to the right-hand side of (6.23) come from the points of S_c where current-carrying conductors pass through, i.e. just terminals A and B. The potential of each conductor is constant and therefore it follows

$$p(t) = \psi_A \int\limits_{S_c} \overline{J} \cdot \overline{n} \, dS + \psi_B \int\limits_{S_c} \overline{J} \cdot \overline{n} \, dS \tag{6.24}$$

If current entering surface S_c is defined as $i = \int_{S_c} \overline{J} \cdot \overline{n} \, dS$, it turns out to be

$$p(t) = (\psi_A - \psi_B) i = v_{AB}(t) i(t) \tag{6.25}$$

i.e. the power entering the element is equal to the product of terminal voltage times current.

Therefore, under the assumption of slowly varying field, Poynting's theorem can be simplified as follows

$$v_{AB} \, i = \frac{dW_m}{dt} + \frac{dW_e}{dt} + p_d \tag{6.26}$$

In general, in a two-terminal element all the three terms at right-hand side are present at a time; in the next section, basic two-terminal elements are defined considering each term as the paramount one.

6.4.1 Resistor Equation

The power entering a resistor is equal to the power loss in the conducting material, i.e. (6.26) becomes

$$v_{AB} \, i = p_d \tag{6.27}$$

From (6.14) and (6.27) it follows

$$v_{AB} i = \int\limits_{\Omega} \sigma E^2 d\Omega \tag{6.28}$$

According to Joule's law [2], the resistance R such that

$$p_d = R i^2 \tag{6.29}$$

can be defined; it follows

$$R = \frac{\int_{\Omega} \sigma E^2 d\Omega}{i^2} = \frac{\int_{\Omega} \sigma E^2 d\Omega}{\left[\int_{\Gamma_c} \overline{J} \cdot \overline{n} \, d\Gamma\right]^2} \tag{6.30}$$

where Γ_c is the resistor cross-section and $\overline{n}$ is the relevant unit vector. The constitutive relation of the resistor

$$v_{AB} = Ri \tag{6.31}$$

follows.

The following remark can be put forward.

Equation (6.30) shows that small errors in computing field strength will not determine a large error in computing resistance: in fact, the averaging effect of volume integral will compensate small errors. In particular, let domain Ω be divided in two complementary sub-domains Ω' and Ω'', respectively; moreover, let the volume of Ω' be much smaller than the volume of Ω'': if the computation of specific power σE^2 is inaccurate in Ω' but accurate in Ω'', the computation of resistance R will be still reliable.

6.4.2 Inductor Equation

The power entering an inductor is equal to the increase in time of the energy stored in the magnetic material, i.e. (6.26) becomes

$$v_{AB}\, i = \frac{dW_m}{dt} \tag{6.32}$$

From (6.14) and (6.32) it follows

$$v_{AB} i = \frac{1}{2}\frac{\partial}{\partial t}\int_{\Omega} \overline{H}\cdot\overline{B}\, d\Omega \tag{6.33}$$

Considering the definition of magnetic vector potential (4.19), it turns out to be

$$\begin{aligned} W_m &= \frac{1}{2}\int_{\Omega} \overline{H}\cdot\overline{B}\, d\Omega = \frac{1}{2}\int_{\Omega} \overline{H}\cdot\left(\overline{\nabla}\times\overline{A}\right) d\Omega \\ &= \frac{1}{2}\int_{\Omega} \overline{\nabla}\cdot\left(\overline{A}\times\overline{H}\right) d\Omega + \frac{1}{2}\int_{\Omega} \left(\overline{\nabla}\times\overline{H}\right)\cdot\overline{A}\, d\Omega \\ &= \frac{1}{2}\int_{\Gamma} \left(\overline{A}\times\overline{H}\right)\cdot\overline{n}\, d\Omega + \frac{1}{2}\int_{\Omega} \overline{J}\cdot\overline{A}\, d\Omega \end{aligned} \tag{6.34}$$

With a suitable choice of integration surface Γ, either $\overline{A}\times\overline{n} = 0$ or $\overline{H}\times\overline{n} = 0$ holds along Γ. Moreover, in a linear material, both current density and magnetic potential are proportional to current i; it follows

$$W_m = \frac{1}{2}\int_{\Omega} \overline{J}\cdot\overline{A}\, d\Omega = \frac{1}{2}Li^2 \tag{6.35}$$

where L is the inductance of the inductor [2], defined as

$$L = \frac{2W_m}{i^2} = \frac{\int_\Omega \overline{J} \cdot \overline{A}\, d\Omega}{\left[\int_{\Gamma_c} \overline{J} \cdot \overline{n}\, d\Gamma\right]^2} = \frac{\int_\Omega \overline{H} \cdot \overline{B}\, d\Omega}{\left[\int_{\Gamma_c} \overline{J} \cdot \overline{n}\, d\Gamma\right]^2} \tag{6.36}$$

where Γ_c is the cross-section of the inductor winding and $\overline{n}$ is the relevant unit vector. The constitutive relation of the inductor

$$v_{AB} = L\frac{di}{dt} \tag{6.37}$$

follows.

The following remark can be put forward.

Equation (6.36) shows that small errors in computing field strength or magnetic induction will not determine a large error in computing inductance: in fact, the average effect of volume integral will compensate small errors. In particular, let domain Ω be divided in two complementary sub-domains Ω' and Ω'', respectively; moreover, let the volume of Ω' be much smaller than the volume of Ω'': if the computation of specific energy $\overline{H} \cdot \overline{B}$ is inaccurate in Ω' but accurate in Ω'', the computation of inductance L will be still reliable.

6.4.3 Capacitor Equation

The power entering a capacitor is equal to the increase in time of the energy stored in the dielectric material, i.e. (6.26) becomes

$$v_{AB}\, i = \frac{dW_e}{dt} \tag{6.38}$$

$$v_{AB} i = \frac{1}{2}\frac{\partial}{\partial t}\int_\Omega \overline{E} \cdot \overline{D}\, d\Omega \tag{6.39}$$

Due to (6.7), (6.17) and (6.8), the following chain of equalities holds

$$\begin{aligned} W_e &= \frac{1}{2}\int_\Omega \overline{E} \cdot \overline{D}\, d\Omega = -\frac{1}{2}\int_\Omega \overline{\nabla}\phi \cdot \overline{D}\, d\Omega \\ &= -\frac{1}{2}\int_\Omega \overline{\nabla} \cdot \phi\overline{D}\, d\Omega + \frac{1}{2}\int_\Omega \phi\overline{\nabla} \cdot \overline{D}\, d\Omega \\ &= \frac{1}{2}\int_{\Gamma_A} \phi\overline{D} \cdot \overline{n}\, d\Omega - \frac{1}{2}\int_{\Gamma_B} \phi\overline{D} \cdot \overline{n}\, d\Omega + \frac{1}{2}\int_\Omega \phi\rho\, d\Omega \end{aligned} \tag{6.39}$$

Considering that the surface charge density on the electrodes is $\rho_s = \overline{D} \cdot \overline{n}$, and $\rho = 0$ holds because there is no free charge in the region between the electrodes, it follows

$$W_e = \frac{1}{2}\phi_A \int_{\Gamma_A} \rho_s \, d\Omega - \frac{1}{2}\phi_B \int_{\Gamma_B} \rho_s \, d\Omega + \frac{1}{2} \int_{\Omega} \phi\rho \, d\Omega$$
$$= \frac{1}{2}\phi_A \int_{\Gamma_A} \rho_s \, d\Omega - \frac{1}{2}\phi_B \int_{\Gamma_B} \rho_s \, d\Omega = \frac{1}{2}(\phi_A - \phi_B)q = \frac{1}{2}v_{AB}q \tag{6.40}$$

In a linear material, charge q is proportional to voltage v_{AB}; therefore, it turns out to be

$$W_e = \frac{1}{2}\int_{\Omega} \overline{E} \cdot \overline{D} \, d\Omega = \frac{1}{2}Cv_{AB}^2 \tag{6.41}$$

where C is the capacitance of the capacitor [2], defined as

$$C = \frac{2W_e}{v_{AB}^2} = \frac{\int_{\Omega} \overline{E} \cdot \overline{D} \, d\Omega}{\left[\int_{\gamma_{AB}} \overline{E} \cdot \overline{t} \, d\gamma\right]^2} \tag{6.42}$$

In (6.42) the integration line can be any open line joining electrode A to electrode B. The constitutive relation of the capacitor

$$i = C\frac{dv_{AB}}{dt} \tag{6.43}$$

follows.

The following remarks can be put forward.

Equation (6.42) shows that small errors in computing field strength or flux density will not determine a large error in computing inductance: in fact, the average effect of volume integral will compensate small errors. In particular, let domain Ω be divided in two complementary sub-domains Ω' and Ω'', respectively; moreover, let the volume of Ω' be much smaller than the volume of Ω'': if the computation of specific energy $\overline{E} \cdot \overline{D}$ is inaccurate in Ω' but accurate in Ω'', the computation of capacitance C will be still reliable.

References

1. Di Barba P, Savini A, Wiak S (2008) Field models in electricity and magnetism. Springer, Berlin, Germany
2. Simonyi K (1963) Foundations of electrical engineering: fields, networks, waves. Pergamon

Chapter 7
Device Miniaturization Principles

7.1 From the Macro- to the Micro-domain

If the size of a mechanical device is reduced by a factor 10, its mass—and, consequently, inertial and gravitational forces—decrease by a factor 10^3; on the other side, forces depending on the device surface like the electrostatic interaction due to surface charge density, decrease by a factor 10^2. Therefore, the ratio of the electrostatic forces to the inertial ones increase by a factor 10; this remark stands as one of the main reasons behind the exploitation of electric field for generating motion in a device exhibiting sub-millimetric size.

7.2 Electric or Magnetic Way to MEMS

It is worth also comparing the electric and magnetic interaction laws. When considering two straight and parallel conductors of infinite length, with current flowing across, according to (4.57) the force exerted by conductor a on an element of conductor b is given by the following expression:

$$d\overline{F}_b = I_b d\overline{\ell}_b \times \overline{B}_a \tag{7.1}$$

where I_b is the current carried by conductor b while $\overline{B}_a$ is the magnetic induction originated by current I_a carried by conductor a. In turn, according to (4.44) the induction magnitude is

$$B_a = \frac{\mu_0 I_a}{2\pi d} \tag{7.2}$$

where d is the distance between the two conductors; the force acts in the direction normal to both conductors [1].

P. Di Barba and S. Wiak, *MEMS: Field Models and Optimal Design*, Lecture Notes in Electrical Engineering 573, https://doi.org/10.1007/978-3-030-21496-8_7

In particular, if the two currents are equal $Ia = Ib = I$ and are equally distributed on the conductor cross-section, the force acting on a section of finite length is

$$F = \frac{\mu_0 I^2 \ell}{2\pi d} = \frac{\pi \mu_0 J^2 \ell}{2d} x^4 \tag{7.3}$$

where J is the current density while x is the radius of the cylindrical conductor.

Now, let a parallel-plate capacitor of infinite length be considered; referring to a square section of side length x, the force acting in the direction normal to the two electrodes is:

$$F = \frac{\sigma^2}{2\varepsilon_0} x^2 \tag{7.4}$$

where σ is the surface charge density on the electrodes.

After comparing (7.3) and (7.4), it can be noted that the magnetic interaction scales down as the fourth power of the geometrical size, while the electric interaction scales as the second power. Therefore, the latter is more convenient in view of the device miniaturization.

Moreover, the absence of current-carrying multi-turn windings, which are space-consuming components, is another convenient feature of electrostatic devices [2].

7.3 Field Strength and Power Density Limits

An obstacle limiting the device operation is the maximum level of voltage that can be applied between two electrodes without causing corona discharge or breakdown effect in the air gap. In this respect, reference is made to the Paschen's law, i.e. the curve that models the breakdown voltage VBD in an ideal gas filling a parallel plate capacitor as a function of the product pd, where p is the gas pressure at constant temperature and d is the gap length. Paschen's curve is shown in Fig. 7.1 for dry air at T = 20 °C.

The curve is characterized by a minimum value corresponding to a critical distance d_{crit}; when $0 < d < d_{crit}$ the slope of the curve is negative and steep. It can be assumed that in this region the distance between the capacitor plates is shorter than the mean free-path of the gas ion and, therefore, the probability of avalanche discharge is very small. Actually, to give origin to the avalanche discharge, a finite path length must be available and a number of subsequent collisions between ions and atoms of the gas must take place before the degenerative phenomenon takes place. If d is very short, the ion is captured by the oppositely-charged capacitor plate without colliding against the neighbouring atoms of gas. As a consequence, it happens that voltages of hundreds of volts can be safely applied to micrometric air gaps at ambient temperature and pressure, largely overcoming the voltage limitation valid in the macroscopic size scale (typically, kilovolts per millimeter).

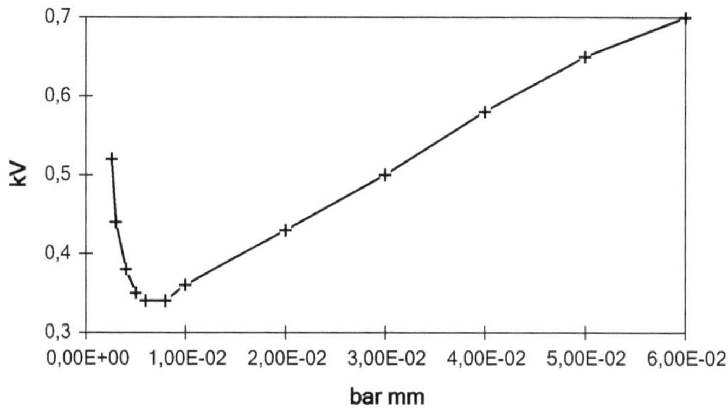

Fig. 7.1 Paschen curve for air at T = 20 °C (parallel-plate capacitor)

In a real-life capacitor, the field in the gas is no longer uniform, due e.g. to the finite size of electrodes and their non-planarity, depending in turn on material properties as well as technological processes. Moreover, impurities entering the air gap can short-circuit energized electrodes and cause breakdown discharge at voltages lower with respect to the theoretical prediction. Nevertheless, the breakdown voltage is still higher than in the macro-domain; simply, the electrode shape might so complex that the basic model in terms of parallel-plate capacitor fails and two- or three-dimensional effects should be taken into account by means of distributed-parameter models.

When a device of rated power is miniaturized, one has an increase in power density which might be substantial. Cooling a miniaturized MEMS in an efficient way can be difficult and therefore the thermal limit is always a severe obstacle to the miniaturization of integrated devices and systems. In this respect, a case study dealing with the power loss minimization is presented in Sects. 14.4.1 and 14.4.2. On the other hand, a high power density with the relevant overheating in a device can be the source of actuation: this is the case of the electro-thermo-elastic actuator in Sects. 13.6 and 13.7.

References

1. Di Barba P, Savini A, Wiak S (2008) Field models in electricity and magnetism. Springer, Berlin, Germany
2. Jones TB, Nenadic NG (2013) Electromechanics and MEMS. Cambridge University Press, Cambridge

Chapter 8
Numerical Methods for Field Analysis of MEMS

8.1 Energy Approach to Finite-Element Analysis

The finite element method has, in recent decades, become by far the most popular technique in computational electromagnetism. Many general purpose computer codes have been developed which provide toolboxes for computer-aided-design (CAD) of devices and systems. The technique is not suitable for hand calculations, and the algorithm is somewhat complicated; nevertheless, the formulation of the simplest two-dimensional case shall be followed to demonstrate the principle and to clarify some aspects of the applications. There are many monographic books on the subject [1, 2, 5, 6].

Consider Laplace's equation in two dimensions for an electrostatic system (see Sect. 3.3)

$$\overline{\nabla}^2 u = \frac{\partial^2 u}{\partial x^2} + \frac{\partial^2 u}{\partial y^2} = 0 \tag{8.1}$$

where the electric field, $\overline{E}$, is given by

$$\overline{E} = -\overline{\nabla} u = -\left(\frac{\partial u}{\partial x}\bar{\iota}_x + \frac{\partial u}{\partial y}\bar{\iota}_y\right) \tag{8.2}$$

A variational principle to the defining equation can be applied by stating that the potential distribution must be such as to minimize the stored field energy. The energy can be expressed as

$$\frac{1}{2}\int_{\Omega} \overline{D}\cdot\overline{E}\, d\Omega = \frac{1}{2}\int_{\Omega} \varepsilon\overline{E}^2\, d\Omega = \frac{1}{2}\int_{\Omega} \varepsilon\left|\overline{\nabla} u\right|^2 d\Omega \tag{8.3}$$

where integration is carried out over the two-dimensional problem region, and is thus taken per unit length [3, 4]. This minimum-energy principle is mathematically

P. Di Barba and S. Wiak, *MEMS: Field Models and Optimal Design*, Lecture Notes in Electrical Engineering 573, https://doi.org/10.1007/978-3-030-21496-8_8

equivalent to the original differential Eq. (8.1) in the sense that a potential distribution which satisfies Laplace's equation will also minimize the energy, and vice versa. An alternative formulation is possible which avoids using energy functionals. It is based on the so-called Galerkin procedure and the method of weighted residuals. The Galerkin procedure is generally easier to apply and leads to a wider class of applications. However, its mathematical formulation is more advanced and it will not be pursued here. For Laplace's equation both formulations give identical results.

8.2 Discretizing the Continuum: Elements and Nodes

Consider a single element in a 2D domain, and assume that the distribution of potential u within the element is adequately approximated by the polynomial expression

$$u = a + bx + cy + dxy + ex^2 + fy^2 + gx^3 + hy^3 + ix^2y + lxy^2 + \cdots \tag{8.4}$$

We choose as many terms in the above equation as there are 'nodes' in the element. In particular, for a rectangle we choose

$$u = a + bx + cy + dxy \tag{8.5}$$

and for a triangle

$$u = a + bx + cy \tag{8.6}$$

In the last case the representation is said to be complete because Eq. (8.6) contains all the terms necessary for a linear variation in two dimensions.

It is easily seen that finite elements offer an immediate extension to higher-order modeling, which makes it possible to increase the accuracy of analysis. For instance, in the case of second-order model the potential distribution within the element is given by the expression

$$u = a + bx + cy + dxy + ex^2 + fy^2 \tag{8.7}$$

Six points (nodes) are needed to find the unknowns: for the purpose the vertices of a triangle as well as its midpoints, which gives rise to a quadratic triangular element, can be chosen. In general, the higher the order the better the accuracy in spite of a higher number of unknowns.

For the sake of simplicity, let the basic case of a linear triangular element be considered. For the three vertices (nodes) of the triangle the potential has the following values

$$u_1 = a + bx_1 + cy_1 \tag{8.8}$$

$$u_2 = a + bx_2 + cy_3 \tag{8.9}$$

$$u_3 = a + bx_3 + cy_3 \tag{8.10}$$

or

$$\begin{bmatrix} u_1 \\ u_2 \\ u_3 \end{bmatrix} = \begin{bmatrix} 1 & x_1 & y_1 \\ 1 & x_2 & y_2 \\ 1 & x_3 & y_3 \end{bmatrix} \begin{bmatrix} a \\ b \\ c \end{bmatrix} \tag{8.11}$$

where (x_1, y_1), (x_2, y_2), and (x_3, y_3) are the coordinates of the vertices; and the determinant of the coefficient matrix in Eq. (8.11) may be recognized as being equal to twice the area of the triangle, A. Rearranging Eq. (8.11) gives

$$\begin{bmatrix} a \\ b \\ c \end{bmatrix} = \begin{bmatrix} 1 & x_1 & y_1 \\ 1 & x_2 & y_2 \\ 1 & x_3 & y_3 \end{bmatrix}^{-1} \begin{bmatrix} u_1 \\ u_2 \\ u_3 \end{bmatrix} \tag{8.12}$$

and substitution back to Eq. (8.6) yields

$$u = \begin{bmatrix} 1 & x & y \end{bmatrix} \begin{bmatrix} 1 & x_1 & y_1 \\ 1 & x_2 & y_2 \\ 1 & x_3 & y_3 \end{bmatrix}^{-1} \begin{bmatrix} u_1 \\ u_2 \\ u_3 \end{bmatrix} \tag{8.13}$$

This last equation may be written as

$$u(x, y) = \sum_{i=1}^{3} u_i \alpha_i(x, y) \tag{8.14}$$

where

$$\alpha_1(x, y) = \frac{1}{2A}\left[(x_2 y_3 - x_3 y_2) + (y_2 - y_3)x + (x_3 - x_2)y\right] \tag{8.15}$$

$$\alpha_2(x, y) = \frac{1}{2A}\left[(x_3 y_1 - x_1 y_3) + (y_3 - y_1)x + (x_1 - x_3)y\right] \tag{8.16}$$

$$\alpha_3(x, y) = \frac{1}{2A}\left[(x_1 y_2 - x_2 y_1) + (y_1 - y_2)x + (x_2 - x_1)y\right] \tag{8.17}$$

At the vertices,

$$\alpha_1(x_1, y_1) = \frac{1}{2A}\left[(x_2 y_3 - x_3 y_2) + (y_2 - y_3)x_1 + (x_3 - x_2)y_1\right] = \frac{2A}{2A} = 1 \tag{8.18}$$

$$\alpha_1(x_2, y_2) = \frac{1}{2A}\left[(x_2y_3 - x_3y_2) + (y_2 - y_3)x_2 + (x_3 - x_2)y_2\right] = 0 \tag{8.19}$$

$$\alpha_1(x_3, y_3) = \frac{1}{2A}\left[(x_2y_3 - x_3y_2) + (y_2 - y_3)x_3 + (x_3 - x_2)y_3\right] = 0 \tag{8.20}$$

and similarly for α_2 and α_3. In general

$$\alpha_i\left(x_j, y_j\right) = 0, \quad i \neq j \tag{8.21}$$

$$\alpha_i\left(x_j, y_j\right) = 1, \quad i = j \tag{8.22}$$

that is, each function vanishes at all vertices but one, where its value is one. Now the energy can be associated with each element, and remembering that in the two-dimensional field the energy will be taken per unit length it may be written

$$W^{(e)} = \frac{1}{2}\int_{\Omega} \varepsilon|\nabla u|^2 d\Omega \tag{8.23}$$

where the integration is performed over the element area. The potential gradient within the element is found from Eq. (8.14) as

$$\nabla u(x, y) = \sum_{i=1}^{3} u_i \nabla\alpha_i(x, y) \tag{8.24}$$

so that the element energy becomes

$$W^{(e)} = \frac{1}{2}\varepsilon\sum_{j=1}^{3}\sum_{i=1}^{3} u_i\left(\int_e \nabla\alpha_i \cdot \nabla\alpha_j d\Omega\right)u_j \tag{8.25}$$

Equation (8.25) may be written in the following compact form

$$W^{(e)} = \frac{1}{2}\varepsilon[U]^T[N]^{(e)}[U] \tag{8.26}$$

where $[U]$ is the column-wise vector or the vertex values of the potential, the superscript T denotes transposition, and the 3×3 square element matrix $[N]$ is defined by

$$N_{i,j}^{(e)} = \int_e \nabla\alpha_i \cdot \nabla\alpha_j \, d\Omega \tag{8.27}$$

For any given triangle, the matrix [N] is readily evaluated. First, the gradients of the α-functions are computed. From Eqs. (8.15), (8.16), and (8.17)

$$\nabla\alpha_1 = \frac{1}{2A}\left[(y_2 - y_3)i_x + (x_3 - x_2)i_y\right] \tag{8.28}$$

$$\nabla\alpha_2 = \frac{1}{2A}\left[(y_3 - y_1)i_x + (x_1 - x_3)i_y\right] \tag{8.29}$$

$$\nabla\alpha_3 = \frac{1}{2A}\left[(y_1 - y_2)i_x + (x_2 - x_1)i_y\right] \tag{8.30}$$

It is interesting to note that for the first-order approximation, given by Eq. (8.6), the gradients of α-functions are constant within an element.

As a consequence, the scalar products of the gradients of α-functions can be easily found. As these gradients are constant within an element, their scalar products will also be constant. Hence integration over the element area will introduce the triangle area, A, as a constant multiplier. The elements of the matrix [N] can now be found, with a typical expression in the form

$$N_{11}^{(e)} = \frac{1}{4A}\left[(y_2 - y_3)^2 + (x_3 - x_2)^2\right] \tag{8.31}$$

$$N_{12}^{(e)} = \frac{1}{4A}\left[(y_2 - y_3)(y_3 - y_1) + (x_3 - x_2)(x_1 - x_3)\right] \tag{8.32}$$

Other entries can be obtained by cyclic permutation of subscripts. This completes the specification for an arbitrary element in the finite element mesh.

The total energy associated with the whole region will be found as the sum of individual element energies

$$W = \sum_e W^{(e)} \tag{8.33}$$

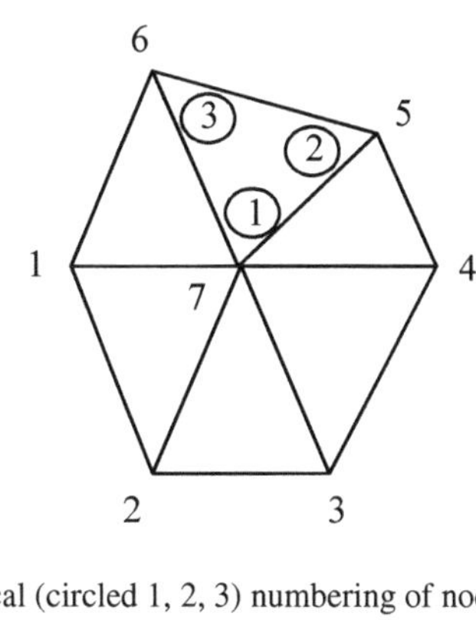

Fig. 8.1 Global (1, 7) and local (circled 1, 2, 3) numbering of nodes

for all elements. When assembling such elements, some nodes will be shared between more than one element as shown in Fig. 8.1, and thus the topology of the actual mesh will directly affect the way in which the global matrix is formulated.

In other words, the global node numbering must be related to the local numbering, and the global matrix must reflect the way in which individual elements are linked to global nodes. This process is discussed in the next section.

8.3 Matrix Assembly

In order to clarify the procedure, a simple example will be used to illustrate the process of assembling the matrix and then solving the system of equations. Consider the region depicted in Fig. 8.2a which models a parallel electrode capacitor. The solution here is trivial: by inspection one has $u_1 = u_2 = 0.5$, and the distribution of the field is uniform.

The general method of solution shall now be applied. The unconstrained nodes are numbered first, and the constrained nodes are numbered last. A particular combination of local and global node numbering schemes is demonstrated in Fig. 8.2b. The data are summarized in Tables 8.1 and 8.2.

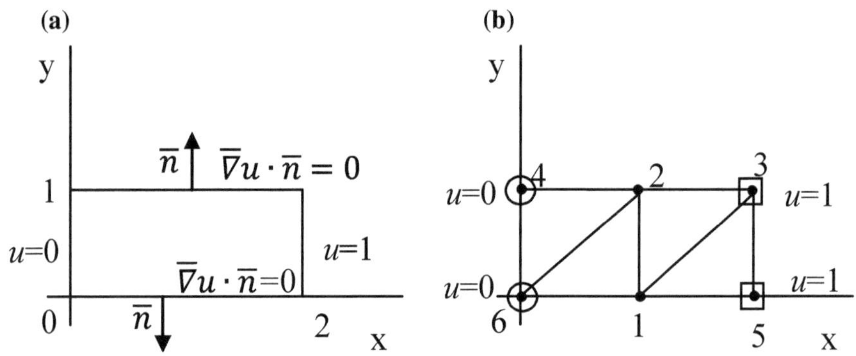

Fig. 8.2 Field region of an ideal parallel-electrode capacitor: **a** boundary conditions, **b** discretizing mesh

Table 8.1 Coordinates and potentials of nodes

Node number	x-coordinate	y-coordinate	Potential
1	1	0	Unknown
2	1	1	Unknown
3	2	1	1
4	0	1	0
5	2	0	1
6	0	0	0

Table 8.2 Elements and node numbers

Element	Vertex 1	Vertex 2	Vertex 3
A	6	1	2
B	6	4	2
C	1	2	3
D	1	5	3

The element matrices may be determined using Eqs. (8.31–8.32). For example, for element A it turns out to be

$$x_3 - x_2 = 0, \quad y_2 - y_3 = -1 \tag{8.34}$$

$$x_1 - x_3 = -1, \quad y_3 - y_1 = 1 \tag{8.35}$$

$$x_2 - x_1 = 1, \quad y_1 - y_2 = 0 \tag{8.36}$$

and the element area $A = 0.5$. Therefore

$$N_{11}^{(A)} = \frac{1}{4A}\left[(y_2 - y_3)^2 + (x_3 - x_2)^2\right] = \frac{1}{2} \tag{8.37}$$

$$N_{12}^{(A)} = \frac{1}{4A}\left[(y_2 - y_3)(y_3 - y_1) + (x_3 - x_2)(x_1 - x_3)\right] = -\frac{1}{2} \tag{8.38}$$

and similarly for the other entries of $[N^{(e)}]$. Inspection of elements B, C, and D shows that, in our example, all the element matrices are the same, that is,

$$[N]^{(A)} = [N]^{(B)} = [N]^{(C)} = [N]^{(D)} = \begin{bmatrix} +\frac{1}{2} & -\frac{1}{2} & 0 \\ -\frac{1}{2} & +1 & -\frac{1}{2} \\ 0 & -\frac{1}{2} & +\frac{1}{2} \end{bmatrix} \tag{8.39}$$

There are six nodes in our mesh, so the global matrix will have dimension of 6×6. Each element matrix will be embedded into the global matrix in a way which depends on the relation between the local vertex numbering and the global node numbering. For example, the elements of the matrix $[N^{(A)}]$ will take the following positions in the global matrix

$$[N]^{(A)} = \begin{bmatrix} N_{66} & N_{61} & N_{62} \\ N_{16} & N_{11} & N_{12} \\ N_{26} & N_{21} & N_{22} \end{bmatrix} \tag{8.40}$$

The subscripts identify where the contribution should be placed in the global matrix. For example, N_{66} is placed in row six, column six of the matrix. Repeating this process for all four elements, and adding appropriate terms, results in the following global matrix

$$[N] = \begin{bmatrix} 2 & -1 & 0 & 0 & -0.5 & -0.5 \\ -1 & 2 & -0.5 & -0.5 & 0 & 0 \\ 0 & -0.5 & 1 & 0 & -0.5 & 0 \\ 0 & -0.5 & 0 & 1 & 0 & -0.5 \\ -0.5 & 0 & -0.5 & 0 & 1 & 0 \\ -0.5 & 0 & 0 & -0.5 & 0 & 1 \end{bmatrix} \tag{8.41}$$

Therefore, for any particular topology and node numbering, the global matrix may be found. This process would be too tedious for hand calculations, but it is easily performed by a computer code.

In order to minimize the total-energy expression, Eq. (8.33) must be differentiated with respect to a typical value of u_k and then equated to zero. Therefore

$$\frac{\partial W}{\partial u_k} = 0 \tag{8.42}$$

where the index k refers to node numbers in the global numbering scheme. In a boundary-value problem, like the problem in the example, some boundary segments have specified potential values. Therefore a subset of the node potentials contained in the vector $[V]$ will assume exactly those prescribed values. To this purpose, the nodes which are free to vary have been numbered first, so leaving all nodes with a prescribed potential to the last. This allows Eq. (8.42) to be rewritten with the matrices in partitioned form,

$$\frac{\partial W}{\partial u_k} = \frac{\partial W}{\partial \left[u_f\right]_k} \left[\left[u_f\right]^T \left[u_p\right]^T \right] \begin{bmatrix} [N_{ff}] & [N_{fp}] \\ [N_{pf}] & [N_{pp}] \end{bmatrix} \begin{bmatrix} [u_f] \\ [u_p] \end{bmatrix} = 0 \tag{8.43}$$

where the subscripts f and p refer to nodes with 'free' and 'prescribed' potentials, respectively. Note that the prescribed potentials cannot vary, and thus differentiation with respect to them is not possible. Accordingly, it turns out to be

$$\frac{\partial W}{\partial u_k} = \frac{\partial W}{\partial \left[u_f\right]_k} \left[\left[u_f\right]^T \left[u_p\right]^T \right] \begin{bmatrix} \lfloor N_{ff} \rfloor u_f + \lfloor N_{fp} \rfloor u_p \\ [N_{pf}] u_f + [N_{pp}] u_p \end{bmatrix} = 0 \tag{8.44}$$

$$\frac{\partial W}{\partial u_k} = \frac{\partial W}{\partial \left[u_f\right]_k} \left[\left[u_f\right]^T \left([N_{ff}] u_f + [N_{fp}] u_p \right) + \left[u_p\right]^T \left([N_{pf}] u_f + [N_{pp}] u_p \right) \right] = 0 \tag{8.45}$$

$$\frac{\partial W}{\partial u_k} = 2[N_{ff}]u_f + [N_{fp}]u_p + [N_{pf}]u_p = 2[N_{ff}]u_f + 2[N_{fp}]u_p = 0 \tag{8.46}$$

All in one, differentiation of energy with respect to the free potentials gives rise to the following matrix equation

$$[[N_{ff}][N_{fp}]]\begin{bmatrix}[u_f]\\ [u_p]\end{bmatrix} = 0 \tag{8.47}$$

and eventually leads to a system of algebraic equations of the form Ax = b; namely

$$[N_{ff}][u_f] = -[N_{fp}][u_p] \tag{8.48}$$

which has a formal solution

$$[u_f] = -[N_{ff}]^{-1}[N_{fp}][u_p] \tag{8.49}$$

In the example

$$[N_{ff}] = \begin{bmatrix} 2 & -1\\ -1 & 2 \end{bmatrix} \tag{8.50}$$

and

$$[N_{fp}] = \begin{bmatrix} 0 & 0 & -\frac{1}{2} & -\frac{1}{2}\\ -\frac{1}{2} & -\frac{1}{2} & 0 & 0 \end{bmatrix} \tag{8.51}$$

so that the final system of equations is given by

$$\begin{bmatrix} 2 & -1\\ -1 & 2 \end{bmatrix}\begin{bmatrix} u_1\\ u_2 \end{bmatrix} = \begin{bmatrix} 0 & 0 & -\frac{1}{2} & -\frac{1}{2}\\ -\frac{1}{2} & -\frac{1}{2} & 0 & 0 \end{bmatrix}\begin{bmatrix} u_3\\ u_4\\ u_5\\ u_6 \end{bmatrix} \tag{8.52}$$

This is of course a simple system of two equations with two unknowns

$$2u_1 - u_2 = 0.5, \quad -u_1 + 2u_2 = 0.5 \tag{8.53}$$

which yields

$$u_1 = u_2 = 0.5 \tag{8.54}$$

on solution, as expected.

A simple notation for the system of equations was used and the nodes had to be numbered in a particular sequence; that is, all potentials free to vary were numbered first, and all potentials with prescribed values were numbered last. In practice, this numbering scheme may not be convenient, and in fact it will not be necessary.

It is also important to note that the finite-element solution is uniquely defined everywhere, not just at the nodes. It is convenient to store the results as a set of nodal potential values, but this is merely a compact representation for the piecewise-planar solution surface which minimizes the system energy.

In conclusion, the finite-element method overcomes the main difficulties of e.g. the finite-difference technique, in particular, accurate matching of irregular boundary shapes and higher-order approximation, and offers more flexibility.

8.4 Solving the System of Equations

The finite-element method transforms the appropriate partial differential equation to a discretized set of algebraic equations $Ax = b$. Boundary conditions and interface conditions between different materials will also be embedded in this equation. Many approaches exist for obtaining the solutions of such equations. They will often be based on a direct method, such as Gaussian elimination, or on an iterative approach, for example, successive over-relaxation (SOR) or conjugate gradients (CGs). The choice of method is important from the practical point of view, as the time and storage requirements of different schemes may be significantly different. The various techniques for solving systems of simultaneous equations are well documented in books on numerical methods. One important property of the equations obtained from finite-element methods is that these equations tend to be very sparse and so does the corresponding matrix of coefficients. Therefore, it is not necessary to store the entries which are zero. Moreover, the matrix is symmetrical about the main diagonal; then, it is sufficient to store the lower (or upper) triangular submatrix.

Most practical codes based on finite elements exploit matrix sparsity in order to keep both computing time and memory requirements within reasonable limits. There are many clever techniques and programming 'tricks' employed to achieve maximum savings. Typically, the nodes are assumed to be randomly numbered initially, and some systematic renumbering scheme is then applied to produce an improved sparsity pattern.

References

1. Chari MVK, Silvester PP (1980) Finite elements in electrical and magnetic field problems. Wiley
2. Davies AJ (1980) The finite element method. Clarendon
3. Di Barba P, Savini A, Wiak S (2008) Field models in electricity and magnetism. Springer, Berlin, Germany
4. Hammond P, Sykulski JK (1994) Engineering electromagnetism. Oxford University Press, Oxford
5. Reece ABJ, Preston TW (2000) Finite element methods in electrical power engineering. Oxford University Press, Oxford
6. Segerlind LJ (1976) Applied finite element analysis. Wiley, New York

Chapter 9
Coupled Fields: Multi-physics Analysis of MEMS

9.1 Multi-physics Analysis of MEMS

When investigating the behaviour of a MEMS device, it might be necessary to model different physical domains which co-exist and interact in the same materials. The relevant analysis implies to solve a system of partial derivative equations, the unknowns of which are time-varying fields acting in the region under study. This is a general concept of coupled-field problem, sometimes referred to as a multiphysics field problem.

Often, analysis and design of MEMS devices ask for solving coupled fields; in this respect, the following remarks can be put forward:

- the *dielectric design* focuses on electrostatic field analysis, with the final aim of avoiding breakdown appearance in all operational conditions;
- the *electro-mechanical design* merges electrostatics, magnetostatics, current flow analysis and elastodynamics to control mechanical deformations caused by electromagnetic forces and torques;
- the *thermal design* joins heat transfer analysis and elastodynamics with the aim of avoiding material degradation caused by thermal overheating.

9.2 Electromagnetism and Coupled Fields

In static conditions, i.e. when fields do not depend on time, single-fields problems have to be modelled: the electrostatic field and the magnetic field co-exist in the same region without interacting if a static distribution of charges and currents is given, respectively. At the model level, the relevant equations can be independently solved [1]. However, if charge and current distributions vary in time, then the electromagnetic field takes place and mutual interactions have to be considered. The time-varying electromagnetic field is governed by the following equations (see Sect. 6.1):

P. Di Barba and S. Wiak, *MEMS: Field Models and Optimal Design*, Lecture Notes in Electrical Engineering 573, https://doi.org/10.1007/978-3-030-21496-8_9

Faraday's equation

$$\overline{\nabla} \times \overline{E} = -\frac{\partial \overline{B}}{\partial t} \tag{9.1}$$

Gauss' electric equation

$$\overline{\nabla} \cdot \overline{D} = \rho \tag{9.2}$$

Ampère's equation

$$\overline{\nabla} \times \overline{H} = \overline{J} + \frac{\partial \overline{D}}{\partial t} \tag{9.3}$$

Gauss' magnetic equation

$$\overline{\nabla} \cdot \overline{B} = 0 \tag{9.4}$$

where scalar function ρ is the charge density [cm^{-3}] and vectors are defined as follows:

$\overline{D}$ electric flux density [cm^{-2}]
$\overline{E}$ electric field strength [V m^{-1}]
$\overline{B}$ magnetic flux density [T]
$\overline{H}$ magnetic field strength [A m^{-1}]
$\overline{J}$ current density [A m^{-2}].

In a three-dimensional domain, Eqs. (9.1)–(9.4) are a set of eight scalar equations which the following constitutive laws

$$\overline{D} = \varepsilon \overline{E} \tag{9.5}$$

$$\overline{B} = \mu \overline{H} + \overline{B}_0 \tag{9.6}$$

$$\overline{J} = \sigma \overline{E} + \overline{J}_0 \tag{9.7}$$

must be added to. In (9.5)–(9.7) a linear material characterized by permittivity ε, permeability μ and conductivity σ, is considered; moreover, in (9.6) the term $\overline{B}_0$ accounts for the permanent magnetization of the magnetic material, if any. In (9.7) the terms $\sigma \overline{E}$ and $\overline{J}_0$ account for eddy current and driving current density, respectively, while in (9.3) the term $\frac{\partial \overline{D}}{\partial t}$ is the displacement current density.

In total, considering both Maxwell equations and constitutive laws, fifteen scalar unknowns (i.e. field components) have to be determined subject to suitable boundary and initial conditions. The system of eight plus nine equations can be solved, because there are two relations among the unknowns which are automatically satisfied. In fact, taking the divergence of (9.1) and the time derivative of (9.4), one obtains an identity. Analogously, taking the divergence of (9.3) and the time derivative of (9.2), the charge continuity equation

$$\overline{\nabla} \cdot \overline{J} + \frac{\partial \rho}{\partial t} = 0 \tag{9.8}$$

follows. In other words, the source terms ρ and $\overline{J}$ are not independent; this makes the coupled-field analysis problem a well-posed one.

9.3 The Electromagnetic Coupled Problem in Terms of Potentials

Despite the field is the quantity of main physical interest, suitable potentials can be introduced to transform vector equations into scalar ones, so reducing the problem complexity [7].

In a simply-connected domain Ω filled in by a linear, homogeneous and isotropic material, after (9.4) the magnetic vector potential $\overline{A}$ [Wb m^{-1}] is defined by the equation

$$\overline{B} = \overline{\nabla} \times \overline{A} \tag{9.9}$$

associated to a suitable gauge condition on $\overline{\nabla} \cdot \overline{A}$ to be specified later on. By means of (9.1) one has

$$\overline{\nabla} \times \left(\overline{E} + \frac{\partial \overline{A}}{\partial t} \right) = 0 \tag{9.10}$$

This means that the vector in brackets can be expressed as the gradient of a scalar potential φ [V] such that

$$\overline{E} + \frac{\partial \overline{A}}{\partial t} = -\overline{\nabla} \varphi \tag{9.11}$$

or

$$\overline{E} = -\overline{\nabla} \varphi - \frac{\partial \overline{A}}{\partial t} \tag{9.12}$$

Substituting (9.12) into (9.3), one obtains

$$\overline{\nabla} \times \overline{H} = \overline{J}_0 - \sigma \overline{\nabla} \varphi - \sigma \frac{\partial \overline{A}}{\partial t} - \varepsilon \frac{\partial}{\partial t} \overline{\nabla} \varphi - \varepsilon \frac{\partial^2 \overline{A}}{\partial t^2} \tag{9.13}$$

Taking (9.9) into account, one has

$$\overline{\nabla} \times \overline{H} = \overline{\nabla} \times \mu^{-1}\overline{\nabla} \times \overline{A} \tag{9.14}$$

Therefore, it turns out to be

$$\overline{\nabla} \times \overline{\nabla} \times \overline{A} + \mu\varepsilon\overline{\nabla}\frac{\partial \varphi}{\partial t} + \mu\varepsilon\frac{\partial^2 \overline{A}}{\partial t^2} = \mu\left(\overline{J}_0 - \sigma\overline{\nabla}\varphi - \sigma\frac{\partial \overline{A}}{\partial t}\right) \tag{9.15}$$

In the case of a current-free and charge-free ideal dielectric region (i.e. $\overline{J}_0 = 0$, $\rho = 0$ and $\sigma = 0$), (9.15) becomes

$$\overline{\nabla} \times \overline{\nabla} \times \overline{A} + \mu\varepsilon\overline{\nabla}\frac{\partial \varphi}{\partial t} + \mu\varepsilon\frac{\partial^2 \overline{A}}{\partial t^2} = 0 \tag{9.16}$$

On the other hand, substituting (9.12) into (9.2) and considering (9.5) gives

$$-\nabla^2\varphi - \frac{\partial}{\partial t}\left(\overline{\nabla} \cdot \overline{A}\right) = 0 \tag{9.17}$$

Equations (9.15) and (9.17) are the link between the two potentials.

Taking into account the vector identity

$$\overline{\nabla} \times \overline{\nabla} \times \overline{A} = \overline{\nabla}\left(\overline{\nabla} \cdot \overline{A}\right) - \overline{\nabla}^2\overline{A} \tag{9.18}$$

and by substituting this expression into (9.15) one has

$$-\overline{\nabla}^2\overline{A} + \overline{\nabla}\left(\overline{\nabla} \cdot \overline{A}\right) + \mu\varepsilon\overline{\nabla}\frac{\partial \varphi}{\partial t} + \mu\varepsilon\frac{\partial^2 \overline{A}}{\partial t^2} = 0 \tag{9.19}$$

or

$$-\overline{\nabla}^2\overline{A} + \overline{\nabla}\left(\overline{\nabla} \cdot \overline{A} + \mu\varepsilon\frac{\partial \varphi}{\partial t}\right) + \mu\varepsilon\frac{\partial^2 \overline{A}}{\partial t^2} = 0 \tag{9.20}$$

The problem of separating (9.29) in two equations, in terms of $\overline{A}$ and φ, respectively, arises.

To this end, if the Lorentz gauge

$$\overline{\nabla} \cdot \overline{A} + \mu\varepsilon\frac{\partial \varphi}{\partial t} = 0 \tag{9.21}$$

is forced, then from (9.20) one obtains

$$\mu\varepsilon\frac{\partial^2\overline{A}}{\partial t^2} - \overline{\nabla}^2\overline{A} = 0 \tag{9.22}$$

which is the wave equation for the magnetic vector potential, subject to appropriate boundary and initial conditions. After determining $\overline{A}$, following (9.21), φ is given by

$$\varphi(x,t) = \varphi_0(x) - \frac{1}{\mu\varepsilon}\int_0^t \overline{\nabla}\cdot\overline{A}(x,t')dt' = 0 \tag{9.23}$$

with $\varphi_0(x)$ to be determined.

Alternatively, forcing gauge (9.21) to Eq. (9.17), the wave equation for the electric scalar potential is obtained

$$\mu\varepsilon\frac{\partial^2\varphi}{\partial t^2} - \nabla^2\varphi = 0 \tag{9.24}$$

After determining φ, $\overline{A}$ can be recovered by means of (9.21).

In the case current with density $\overline{J}_0$ and charge with density ρ are present, the two above equations become

$$\mu\varepsilon\frac{\partial^2\overline{A}}{\partial t^2} - \overline{\nabla}^2\overline{A} = \mu\overline{J}_0 \tag{9.25}$$

and

$$\mu\varepsilon\frac{\partial^2\varphi}{\partial t^2} - \nabla^2\varphi = \frac{\rho}{\varepsilon} \tag{9.26}$$

respectively.

It can be remarked that, thanks to the Lorentz gauge (9.21), $\overline{A}$ depends only on $\overline{J}_0$ while φ depends only on ρ.

In the case of a conductor ($\rho = 0$, $\sigma \neq 0$), by forcing the following gauge

$$\overline{\nabla}\cdot\overline{A} + \mu\varepsilon\frac{\partial\varphi}{\partial t} + \mu\sigma\varphi = 0 \tag{9.27}$$

from (9.15) and (9.18) it follows

$$-\overline{\nabla}^2\overline{A} + \overline{\nabla}\left(\overline{\nabla}\cdot\overline{A} + \mu\varepsilon\frac{\partial\varphi}{\partial t} + \mu\sigma\varphi\right) + \mu\varepsilon\frac{\partial^2\overline{A}}{\partial t^2} + \mu\sigma\frac{\partial\overline{A}}{\partial t} = \mu\overline{J}_0 \tag{9.28}$$

or

$$\mu\varepsilon\frac{\partial^2\overline{A}}{\partial t^2}+\mu\sigma\frac{\partial\overline{A}}{\partial t}-\overline{\nabla}^2\overline{A}=\mu\overline{J}_o \tag{9.29}$$

After determining $\overline{A}$ and so $\overline{\nabla}\cdot\overline{A}$, φ can be recovered solving (9.27) with initial condition $\varphi_0(x)$.

The special case of the electromagnetic field in a conductor is particularly important in low-frequency applications (eddy current problem). In fact, following the $\overline{A}-\varphi$ formulation, a magnetic vector potential $\overline{A}$ [Wb m^{-1}] is introduced by (9.9); moreover, an electric scalar potential φ [V] is defined according to (9.11): this way, fields $\overline{E}$ and $\overline{H}$ are expressed by means of two potentials, namely $\overline{A}$ and φ.

At this point in time, the approximation of quasi-static condition is necessary: if the term $\overline{J}_0$ in (9.7) exhibits time-harmonic variations, their period is supposed to be much higher than the time constant $\sigma^{-1}\varepsilon$ of the material. Then, displacement current density $\frac{\partial\overline{D}}{\partial t}$ in (9.3) may be neglected with respect to driving $\overline{J}_0$ and induced $\sigma\overline{E}$ current densities; (9.3) is simplified as

$$\overline{\nabla}\times\overline{H}=\sigma\overline{E}+\overline{J}_0 \tag{9.30}$$

Now, to determine potentials, from (9.30), taking into account (9.9) and (9.12), it turns out to be

$$\overline{\nabla}\times\overline{\nabla}\times\overline{A}=\mu\overline{J}_0-\mu\sigma\frac{\partial\overline{A}}{\partial t}-\mu\sigma\overline{\nabla}\varphi \tag{9.31}$$

In order to specify $\overline{A}$ uniquely, a further condition must be introduced. This may be the following gauge

$$\overline{\nabla}\cdot\overline{A}+\mu\sigma\varphi=0 \tag{9.32}$$

which corresponds to (9.27) if the time derivative of φ is disregarded.

Taking (9.18) into account, from (9.31) and (9.32) one obtains

$$\mu\sigma\frac{\partial\overline{A}}{\partial t}-\overline{\nabla}^2\overline{A}+\overline{\nabla}\left(\overline{\nabla}\cdot\overline{A}+\mu\sigma\varphi\right)=\mu\sigma\frac{\partial\overline{A}}{\partial t}-\overline{\nabla}^2\overline{A}=\mu\overline{J}_0 \tag{9.33}$$

The latter is the diffusion equation in terms of vector potential, subject to appropriate boundary and initial conditions; it is an approximation of Eq. (9.29) in the quasi-static state. After determining $\overline{A}$, scalar potential $\varphi=-(\mu\sigma)^{-1}\,\overline{\nabla}\cdot\overline{A}$ is recovered.

Finally, in the static case, (9.29) and (9.33) reduce to

$$-\overline{\nabla}^2\overline{A}=\mu\overline{J}_0 \tag{9.34}$$

i.e. the Poisson equation of magnetostatics (single-field magnetic problem).

In turn, (9.26) reduces to

$$-\nabla^2\varphi = \frac{\rho}{\varepsilon} \tag{9.35}$$

i.e. the Poisson equation of electrostatics (single-field electric problem).

9.4 The Electro-Thermal Coupled Field

The electro-thermal case is a typical coupled-field problem: a relevant case study is here referred to.

Let a cylindrical workpiece, coaxially located inside a multi-turn cylindrical winding with rectangular cross section, be considered; an axial-symmetry problem is so originated. In the winding an AC current uniformly distributed with density *Jo* takes place: due to electromagnetic induction principle (Faraday law, (9.1)), a current is induced in the workpiece. Finally, based on the Joule effect, the associated power loss gives rise to a temperature increase of the workpiece [2, 4, 5].

In general, however, both electrical and thermal properties of the workpiece vary with temperature: this means that the induced current makes the temperature increase, which in turn modifies the values of electrical conductivity and magnetic permeability, eventually changing the current distribution. As a consequence, the magnetic problem and the thermal one cannot be independently studied [3].

Under quasi-static conditions, the magnetic problem is ruled by the diffusion equation of magnetic vector potential (9.33) subject to suitable boundary and initial conditions; (9.33) can be rewritten as

$$-\nabla^2\overline{A} = \mu\overline{J}_0 - \mu\sigma\frac{\partial\overline{A}}{\partial t} \tag{9.36}$$

where, at the right-hand side, the eddy current $\sigma\frac{\partial\overline{A}}{\partial t}$, opposing the driving current $\overline{J}_0$, appears. In particular, the induced electric field, responsible of the eddy current circulation, is

$$\overline{E} = \frac{\partial\overline{A}}{\partial t} \tag{9.37}$$

Due to the AC regime of current and so magnetic field, (9.36) can be transformed in the frequency domain; it turns out to be

$$-\nabla^2\dot{\overline{A}} + j\omega\mu\sigma\dot{\overline{A}} = \mu\dot{\overline{J}}_0 \tag{9.38}$$

where $\dot{\overline{\mathrm{A}}}$ and $\dot{\overline{\mathrm{J}}}_0$ are the complex vectors corresponding to $\overline{\mathrm{A}}$ and $\overline{\mathrm{J}}_0$, respectively, subject to suitable boundary conditions. The induced electric field (9.37) transforms as

$$\dot{\overline{E}} = j\omega\dot{\overline{A}} \tag{9.39}$$

In turn, the thermal problem is a transient one, ruled by the Fourier's equation of temperature T

$$-\overline{\nabla} \cdot k\overline{\nabla}T + c\rho\frac{\partial T}{\partial t} = p \tag{9.40}$$

with

k thermal conductivity [W m^{-1} K^{-1}]
c specific heat [J kg^{-1} K^{-1}]
ρ mass density [kg m^{-3}]
p power density [W m^{-3}]

subject to boundary conditions

$$-k\frac{\partial T}{\partial n} = h(T - T_e) + \epsilon\sigma\left(T^4 - T_e^4\right) \tag{9.41}$$

with n external normal to the boundary, h convection coefficient, T_e environmental temperature, ε surface emissivity, and $\sigma = 506{,}704 \times 10^{-8}$ [W m^{-2} K^{-4}] the Stefan constant, as well as to initial conditions $T(x, t_0) = T_0(x)$, respectively.

The power density due to induced current in the workpiece is

$$p_1 = \sigma\left\|\overline{E}\right\|^2 \tag{9.42}$$

while the one in the winding is

$$p_2 = \sigma^{-1}\left\|\overline{J}_0 + \sigma\overline{E}\right\|^2 \tag{9.43}$$

In the winding, in fact, both driving current and induced current take place. It can be remarked that the total instantaneous loss in the winding is given by

$$P_2 = \int_{\Omega}\left(\sigma^{-1}\left\|\overline{J}_0\right\|^2 + \sigma\left\|\overline{E}\right\|^2 + 2\left\|\overline{J}_0\right\|\left\|\overline{E}\right\|\right)d\Omega \tag{9.44}$$

Now,

$$\int_{\Omega}\left\|\overline{J}_0\right\|\left\|\overline{E}\right\|d\Omega = 0 \tag{9.45}$$

because J_0 is uniform and σE is an eddy current and therefore has no net flow across the section Ω. As a consequence, the active specific loss is given by

$$p_2' = \sigma^{-1} \left\| \overline{J}_0 \right\|^2 + \sigma \left\| \overline{E} \right\|^2 \tag{9.46}$$

As far as the time advance of (9.40), the following remark can be put forward: in materials thermal time constants are much longer the electrical ones; consequently, the magnetic problem governed by (9.38) is a steady-state one, while the thermal problem governed by (9.40) is actually a transient one.

Actually, (9.38) and (9.40) model the coupled field problem here considered. If material properties like e.g. magnetic permeability or electric conductivity are considered to be constant, then (9.38) can be solved independently from (9.40) and the problem is said to be weakly coupled: solve magnetic equation, first, and thermal equation, next. This gives rise to a cascade scheme of solution, because the right-hand side of (9.40) is sufficiently computed through (9.39) and (9.42) knowing the solution of (9.38).

In contrast, when e.g. the temperature dependence of magnetic permeability $\mu = \mu(T)$ or electric conductivity $\sigma = \sigma(T)$ is taken into account, then the problem is said to be strongly coupled, because the coefficients of (9.38) depend on the unknown of (9.40) [4, 6]. This is the case, for instance, when material temperature exceeds its Curie temperature and the material losses its ferromagnetic property behaving as a paramagnetic one; likewise, conductor conductivity decreases with temperature.

For the sake of simplicity, let time-derivative of temperature be zero i.e. thermal regime is assumed; therefore, the thermal equations becomes

$$-\overline{\nabla} \cdot k \overline{\nabla} T = \sigma^{-1} \left\| \dot{\overline{J}}_0 \right\|^2 + \sigma \omega^2 \left\| \dot{\overline{A}} \right\|^2 \tag{9.47}$$

It can be noted that the first term at the right-hand side models Joule's losses, while the second term models eddy current losses. Moreover, the quadratic dependence on vector potential can be linearized based on truncated Taylor's expansion. The following formulation follows:

$$-\nabla^2 \dot{\overline{A}} + j\omega\mu\sigma \dot{\overline{A}} = \mu \dot{\overline{J}}_0 \tag{9.48}$$

$$-\overline{\nabla} \cdot k \overline{\nabla} T - \left(2\sigma\omega^2 \left\| \dot{\overline{A}} \right\| \right) \left\| \dot{\overline{A}} \right\| = \sigma^{-1} \left\| \dot{\overline{J}}_0 \right\|^2 \tag{9.49}$$

The system can be solved by means of finite-element analysis: each node of the discretization grid is attributed four unknowns: one for temperature and three for magnetic potential. In the two-dimensional case or axial-symmetric case, the magnetic vector potential is a scalar quantity; so, only two unknowns, i.e. temperature and potential, are defined per each node. In general, however, magnetic domain and thermal domain are different: for instance, in the case study the magnetic domain incorporates workpiece, winding and surrounding dielectric material, while the thermal domain is just the workpiece. Consequently, the relevant grids are also different; for the sake of generality, nm magnetic nodes and nt thermal nodes can be assumed.

Then, (9.48)–(9.49) can be cast in matrix form

$$\begin{bmatrix} K_A & 0 \\ C & K_T \end{bmatrix} \begin{bmatrix} A \\ T \end{bmatrix} = \begin{bmatrix} F_A \\ F_T \end{bmatrix} \tag{9.50}$$

where A and T are the unknown column vectors of size $(n_m, 1)$ and $(n_t, 1)$ respectively. The block matrix at the left-hand side is composed of the following terms:

$$K_A = \left[-\nabla^2 + j\omega\mu\sigma\right] \tag{9.51}$$

i.e. the (n_m, n_m) magnetic operator matrix stemming from finite-element approximation,

$$K_T = \left[-k\nabla^2\right] \tag{9.52}$$

i.e. the (n_t, n_t) thermal operator matrix stemming from finite-element approximation, and the (n_t, n_m) coupling matrix C, which in the special case of a triangular element of the first order is

$$C^{(e)} = 2\sigma^{(e)}\omega^2 \begin{bmatrix} A_1^{(e)} & 0 & 0 \\ 0 & A_2^{(e)} & 0 \\ 0 & 0 & A_3^{(e)} \end{bmatrix} \tag{9.53}$$

with e element index and A_k, $k = 1, 3$ nodal potentials.

It can be noted that entry (1, 2) in block matrix (9.50) is a null (n_m, n_t) matrix, because in the first equation the unknown temperature does not appear. Moreover, the entries of C matrix depend on the magnetic unknown, and this makes the system of coupled equations a non-linear one; they depend also on the driving term J_0 through angle frequency ω.

In turn, data column vectors are

$$F_A = \mu J_0 \tag{9.54}$$

i.e. the $(n_m, 1)$ magnetic driving term, and

$$F_T = \sigma^{-1} J_0^2 \tag{9.55}$$

i.e. the $(n_t, 1)$ thermal driving term.

The following remarks can be put forward.

In DC conditions, $\omega = 0$ holds and the coupling matrix C is null. In this case, the system matrix is a block-diagonal one and the two equations can be independently solved (weakly coupled problem).

In AC conditions, $\omega \neq 0$ holds and the coupling matrix C is not null. In this case, the two equations cannot be independently solved (strongly coupled problem).

9.5 Subsequent Substitution Algorithm

To this end, the Banach-Picard algorithm of subsequent substitutions can be used. Assuming that magnetic grid and thermal grid are the same ($n_m = n_t$), indicating k as the iteration index and i as the node index, a possible pseudo-code is the following:

(i) the magnetic equation is solved $A_i^{(k+1)} = \left[K_A^{(k)}\right]^{-1} F_A^{(k)}$;
(ii) the magnetic contribution to thermal source is updated $C^{(k)} A_i^{(k+1)}$;
(iii) the thermal equation is solved $T_i^{(k+1)} = \left[K_T^{(k)}\right]^{-1} \left[F_T^{(k)} - C^{(k)} A_i^{(k+1)}\right]$;
(iv) the twofold convergence test

$$\frac{\left\| A_i^{(k+1)} - A_i^{(k)} \right\|}{\left\| A_i^{(k)} \right\|} < \varepsilon_A$$

and

$$\frac{\left\| T_i^{(k+1)} - T_i^{(k)} \right\|}{\left\| T_i^{(k)} \right\|} < \varepsilon_T$$

is performed, with ε_A and ε_T prescribed thresholds of accuracy for magnetic potential and temperature, respectively.

References

1. Di Barba P, Savini A, Wiak S (2008) Field models in electricity and magnetism. Springer, Berlin, Germany
2. Di Barba P, Dolezel I, Karban P, Kus P, Mach F, Mognaschi ME, Savini A (2014a) Multiphysics field analysis and multiobjective design optimization: a benchmark problem. Inverse Prob Sci Eng 22(7):1214–1225
3. Di Barba P, Dolezel I, Mognaschi ME, Savini A, Karban P (2014b) Non-linear multi-physics analysis and multi-objective optimization in electroheating applications. IEEE Trans Magn 50(2)
4. Di Barba P, Mognaschi ME, Lowther DA, Dughiero F, Forzan M, Lupi S, Sieni E (2017) A benchmark problem of induction heating analysis. IJAEM 53(S1):S139–S149
5. Di Barba P, Dughiero F, Forzan M, Mognaschi ME, Sieni E (2018a) New solutions to a multi-objective benchmark problem of induction heating: an application of computational biogeography and evolutionary algorithms. Arch Electr Eng 67(1):139–149

6. Di Barba P, Mognaschi ME, Bullo M, Dughiero F, Forzan M, Lupi S, Sieni E (2018b) Field models of induction heating for industrial applications. Przegl Elektrotech 94(3):1–5
7. Simonyi K (1963) Foundations of electrical engineering: fields, networks, waves. Pergamon, New York

Chapter 10
Numerical Methods for MEMS Design: Inverse Problems

10.1 Direct and Inverse Problems

In engineering science, direct problems are defined as those where, given the input or the cause of a phenomenon or of a process in a device, the purpose is that of finding the output or the effect.

Inverse problems, conversely, are those where, given the measured or expected output or effect, one wants to determine the input or the cause; moreover, inverse problems are also those where, given the input and the corresponding output, one tries to understand their interconnection.

The two types of problems, when applied to the same phenomenon or process, represent the two logical ways of conceiving it: from input to output or the other way round. The latter viewpoint is central for design.

In electromagnetics, inverse problems may appear in either of two forms:

- given measured data, which may be affected by noise or error, in a field region, to identify or recover the relevant field sources or material properties or boundary conditions of the region (identification or parameter-estimation problems);
- given desired fields in a device, or given the device performance based on them, to determine, or design, sources or materials or shape of the device, producing the specified performance (synthesis or optimal design problems).

In particular, optimal shape design problems, which are very popular in all branches of engineering, belong to a group of inverse problems where the purpose is to find the geometry of a device which can provide a prescribed behaviour or an optimal performance [2].

If optimal design problems should be solved only by means of a trial-and-error approach, it would not be possible to know something a priori about their solution, which in turn would rely just on the designer experience and intuition. On the contrary, the study of inverse problems puts the ground for a systematic approach to the design. This line will be developed throughout the book.

P. Di Barba and S. Wiak, *MEMS: Field Models and Optimal Design*, Lecture Notes in Electrical Engineering 573, https://doi.org/10.1007/978-3-030-21496-8_10

Inverse problems have enormously influenced the development of the natural sciences, a fact not generally appreciated. Often, such problems involve determining physical laws e.g. through indirect observations. A famous example is the Newton derivation of the inverse-square law for planet motions: given the form of the orbit, derive the form of the force law that would generate the given orbit.

Only the inverse-square law can explain the motion along elliptically shaped orbits: so doing, he solved the inverse Kepler problem. Newton organised his results in the *Principia Mathematica Philosophiae Naturalis* (1687), where he presents the solution just to many direct and inverse problems for planet orbits.

Two centuries later, Maxwell, in *A treatise of Electricity and Magnetism* (1892), for the first time put forward a problem of optimal shape design of an air-cored inductor: given a finite piece of wire, wind it in such a way to realize an inductor of rectangular cross-section having maximum inductance.

The solution, found by Maxwell himself, is particularly simple: the optimal winding exhibits a squared cross-section such that $2r = 3l$, where r and l are mean radius of the winding and side length of the square, respectively.

10.2 Insidiousness of Inverse Problems

Despite the conceptual importance of inverse problems, the greater part of engineering science is dominated by direct problems, i.e. problems that can be characterized as those in which exactly enough information is provided to carry out a unique solution. A general description of direct problems may be given as follows:

let x, y and A symbolise the input, the output and the operator modelling the input-to-output transformation, respectively. Then, the direct problem is to find Ax, i.e. the value of the given operator at a point in its domain.

Conversely, assuming that the operator A is invertible, the inverse problem for A is the direct problem for A^{-1}; therefore, if A is not invertible, the solution to the inverse problem does not exist.

On the other hand, if operator A represents a function, then for any given input x in its domain, a unique output y is determined: in other words, the direct problem has a unique solution. There is no guarantee, however, that the inverse problem $A^{-1}y$ has a unique solution: in fact, for the same given y, $x_1 = A^{-1}y$ might be different from $x_2 = A^{-1}y$.

Moreover, if the operator A is continuous in some sense, then the solution to the direct problem is stable with respect to small changes in the input, i.e. $\frac{dy}{y}$ is small if $\frac{dx}{x}$ is small. Even when the operator has a well-defined inverse A^{-1}, so that the inverse problem is uniquely solvable, there is no guarantee that its solution is stable against small changes dy; the inverse operator might, in fact, be discontinuous [4].

From the mathematical viewpoint, following the Hadamard definition [5, 7], well-posed problems (or properly, correctly posed problems) are those for which:

(i) a solution always exists;
(ii) there is only one solution;
(iii) a small change of data leads to a small change in the solution.

The last property implies that the solution depends continuously upon the data, which often are measured quantities and therefore are affected by noise or error.

Ill-posed problems, instead, are those for which:

(i) a solution may not exist;
(ii) there may be more than one solution;
(iii) a small change of data may lead to a big change in the solution.

For the sake of an example, to focus just on the dependence of the solution on data, let the differentiation of a function *g(x)*, known approximately, be considered. Formally, the superposition $g(x) = \tilde{g}(x) + \lambda \sin\left(\frac{x}{\lambda}\right)$, where $\tilde{g}(x)$ is the exact information to recover and $\lambda \sin\left(\frac{x}{\lambda}\right)$ is an error term with zero mean on a period $\lambda > 0$, holds. It is straightforward to see that $g(x) \to \tilde{g}(x)$ if $\lambda \to 0$; however, after differentiation, one has $\frac{dg}{dx} = \frac{d\tilde{g}}{dx} + \cos\left(\frac{x}{\lambda}\right)$. Apparently, the condition $\lambda \to 0$ does not imply the condition $\frac{dg}{dx} \to \frac{d\tilde{g}}{dx}$. This shows that arbitrarily small errors in data—i.e. function *g(x)*—can determine uncontrolled errors in results, *i.e.* the derivative of *g(x)*; in fact, one has: $\inf_x \frac{d\tilde{g}}{dx} - 1 \leq \frac{dg}{dx} \leq \sup_x \frac{d\tilde{g}}{dx} + 1$.

About ill-posed problems, the following remark can be put forward.

Identification problems have always a solution at least, while a solution may not exist for optimal design problems; this happens when e.g. the prescribed quantity does not fit with data. On the contrary, if multiple solutions exist to a given problem, they might be similar, differing by e.g. a degree of smoothness or exactness.

All these reasons make inverse problems insidious; therefore, breaking the cultural 'tyranny' of direct problems in science is hardly difficult.

10.3 Classification of Inverse Problems

There are many ways to classify inverse problems. The formulation of inverse problems in electricity and magnetism implies to associate a procedure for field computation (direct problem) and a procedure for the solution of the inverse problem. Therefore, a classification can be based on the approach for field computation (e.g. integral or differential, analytical or numerical). A more satisfactory classification can be made, according to the formulation of the inverse problem and the relevant mathematical method employed for its solution; this viewpoint will be developed later.

When the given data come from measurements and the parameters governing field equations, including material properties, are to be found, one speaks of identification problems.

Otherwise, when the given data are arbitrarily taken and the field source or specifications of the field region (e.g. boundary conditions) are required, the problem is called a synthesis problem.

In engineering applications, often, the goal is to design the geometry of a device so that a prescribed performance of the device, depending on the field, is obtained. This kind of problem is commonly defined as optimal shape design problem.

The ultimate goal of the problem is to perform an automated optimal design (AOD), when the solution is obtained automatically in terms of the required or best performance.

In Chap. 14, some examples of optimal shape design problems solved by means of automated procedures are presented and discussed.

10.4 Solving Inverse Problems by Means of Rectangular Systems of Equations

In general, the numerical solution of field problems leads to a system of algebraic equations of the type:

$$Ax = b \tag{10.1}$$

where A is a full-rank rectangular $m \times n$ matrix, x is the unknown n-vector and b the known m-vector.

If $m < n$, the system is called under-determined. If, on the contrary, $m > n$, the system is called over-determined; the latter case is the most frequent when dealing with inverse problems, because one normally has more conditions to fulfil than degrees of freedom available.

Finally, if $m = n$, the matrix A is square. In this case, if $\det(A) \neq 0$, then A is non-singular; therefore A^{-1} exists and the corresponding system of equations has a unique solution for any b. This is the typical case when dealing with direct problems.

As far as the effect of a small perturbation of b on x is concerned, the following remark can be put forward. Supposing $m = n$, let the condition number of A be defined as follows:

$$\text{cond}(A) \equiv \|A\| \left\|A^{-1}\right\| = \frac{\lambda_{\text{max}}}{\lambda_{\text{min}}} \geq 1 \tag{10.2}$$

where λ_{max} and $\lambda_{\text{min}} \neq 0$ are maximum and minimum eigenvalue of matrix A, respectively. If cond(A) is large, then the matrix is said to be ill-conditioned and the solution might be perturbed substantially by even a small change of b.

If A is rectangular, theoretically the inverse of A does not exist and the system of equations has no or infinite solutions. However, if $m > n$ and the rank of A is equal to n (i.e. the n columns of A are linearly independent), a pseudo-inverse of A can be looked for, by means of suitable numerical techniques like least-squares or singular-value decomposition.

10.5 Least Squares

If A is a $m \times n$ matrix ($m > n$) of rank n and b is a given m-vector, then a solution to (10.1) can be found by minimising a norm, for instance the Euclidean or two-norm, of the residual Ax-b. The latter is defined as

$$r(x) = \|Ax - b\|_2^2 = x^T A^T Ax - 2x^T A^T b + b^T b \tag{10.3}$$

The gradient of the residual is

$$\overline{\nabla} r(x) = 2A^T Ax - 2A^T b \tag{10.4}$$

Apparently, the residual has a unique minimum point $\tilde{x}$ such that $\overline{\nabla} r(\tilde{x}) = 0$. The so-called normal equations associated to (10.1) are obtained forcing just the latter condition, giving

$$2A^T Ax - 2A^T b = 0 \tag{10.5}$$

and, therefore,

$$A^T Ax = A^T b \tag{10.6}$$

where $A^T A$ is a square $n \times n$ matrix. It can be proven that the vector

$$\tilde{x} = (A^T A)^{-1} A^T b \tag{10.7}$$

fulfils the condition

$$\|A\tilde{x} - b\|_2 \leq \|Ax - b\|_2 \tag{10.8}$$

for each n-dimensional vector x and so $\tilde{x}$ is the least-square solution to (10.1); matrix $(A^T A)^{-1} A^T$ is called pseudo-inverse of A.

In principle, if A has full-column rank, $A^T A$ is positive definite; however, from the numerical viewpoint, solving (10.6) might fail for a twofold reason:

- the magnification of ill-conditioning when passing from A to $A^T A$, resulting in cond$(A^T A) \gg 1$;
- the round-off errors after calculating the entries of $A^T A$.

Therefore, the use of normal equations is not recommended because it might lead to instability and inaccuracy.

10.6 Singular-Value Decomposition

A more effective approach is given e.g. by the Singular Value Decomposition (SVD) method; basically, it consists of decomposing the matrix A, which is assumed to be full-column rank ($m > n$), into the product of three matrices, i.e. a $m \times m$ orthogonal matrix U, a $m \times n$ block diagonal matrix S, a $n \times n$ orthogonal matrix V, such that $A = USV^T$.

In particular, it results

$$S = \begin{bmatrix} \Sigma & 0 \\ 0 & 0 \end{bmatrix} \tag{10.9}$$

with $\Sigma = \mathrm{diag}(\sigma_1, \ldots, \sigma_n)$. The diagonal entries of Σ are the singular values of A.

The solution to the least-square problem is then given by

$$x^* = V\, S^{-1} U^T b \tag{10.10}$$

with

$$S^{-1} = \begin{bmatrix} \Sigma^{-1} & 0 \\ 0 & 0 \end{bmatrix} \tag{10.11}$$

and $\Sigma^{-1} = \mathrm{diag}\left(\sigma_1^{-1}, \ldots, \sigma_n^{-1}\right)$. Also matrix $VS^{-1}U^T$ is a pseudo-inverse of A.

10.7 Regularization

In field theory, using an integral approach, equations of the type:

$$g(x) = \int_{\Omega_1} K(x, y) f(y) dy, \quad x \in \Omega_0 \subseteq \Re^3, \quad y \in \Omega_1 \subseteq \Re^3 \tag{10.12}$$

where Ω_0 is the field domain and Ω_1 is the source domain, are frequently dealt with. When f is given, g is the unknown and K is the known kernel; the problem of finding g is a direct problem. In electricity and magnetism, (10.12) is nothing but the Green formula relating potential g to source density f in an unbounded domain [3].

In turn, when g is given, f is the unknown and K is the known kernel; the problem of finding f is an inverse problem. In this case, (10.12) is called Fredholm equation of first kind [7].

The regularization method was proposed as a way to stabilize the solution of the Fredholm equation of the first kind.

For this class of integral equations, the solution *f(y)* does not depend continuously on the given function *g(x)*. Since the output is not stable against small perturbation of the input, problem (10.12) violates the Hadamard conditions of well-posedness (see Sect. 10.2).

In (10.12) the integral operator $\int K(x, y) \cdot dy$ can be discretized by means of a finite-difference grid composed of n nodes, while the known term *g(x)* can be discretized on another grid of $m > n$ nodes. This gives rise to matrix A and vector b, approximating the integral operator and the known term, respectively.

Due to the ill-posedness of the continuous problem (10.12), also the discretized problem, i.e. the resulting set of linear algebraic Eq. (10.1), where vector x approximates function *f(y)*, is ill-posed. Nonetheless, taking into account some a priori information about the solution, it is possible to convert (10.12) into a well-posed problem: for instance, if the norm of the solution x should be bounded, it makes sense to incorporate a penalty term into the problem formulation.

To this end, let the Tikhonov functional be considered [8]; it is defined as

$$\begin{aligned} T(x_\alpha) &\equiv \|Ax_\alpha - b\|_2^2 + \alpha\|x_\alpha\|_2^2 \\ &= x_\alpha^T A^T A x_\alpha - 2x_\alpha^T A^T b + b^T b + \alpha x_\alpha^T x_\alpha \end{aligned} \tag{10.13}$$

where regularized solution x_α depends on parameter $\alpha \in (0, 1)$.

Then, the regularization problem reads

$$\text{find} \inf_{x_\alpha \in X} T(x_\alpha) \tag{10.14}$$

By forcing the equilibrium condition $\overline{\nabla} T(x_\alpha) = 0$, one finds a unique minimum point; in fact, one has

$$\overline{\nabla}\left[\|Ax_\alpha - b\|_2^2 + \alpha\|x_\alpha\|_2^2\right] = 2A^T A x_\alpha - 2A^T b + 2\alpha x_\alpha = 0 \tag{10.15}$$

The solution $\tilde{x}_\alpha$ to (10.15) is the so-called quasi-solution to problem (10.12). Therefore, $\tilde{x}_\alpha$ solves the system of linear equations $\alpha x_\alpha + A^T A x_\alpha = A^T b$, or, equivalently,

$$\left(A^T A + \alpha I\right) x_\alpha = A^T b \tag{10.16}$$

The latter is the Euler equation associated to Tikhonov functional. If columns of the augmented matrix $A^T A + \alpha I$ are linearly independent, then the solution $\tilde{x}_\alpha$ is unique and it can be proven that it depends continuously on $A^T b$. If an a priori estimate x_0 of the quasi-solution $\tilde{x}_\alpha$ is known, then the second contribution to Tikhonov functional is updated as $\alpha\|x_\alpha - x_0\|_2^2$ and (10.15) becomes

$$\left(A^T A + \alpha I\right) x_\alpha = A^T b + \alpha x_0 \tag{10.17}$$

In other words, $\tilde{x}_\alpha$ keeps the residual $\|Ax_\alpha - b\|_2^2$ small in a stable way, which is controlled by the penalty term $\alpha\|x_\alpha\|_2^2$. As far as numerical aspects are concerned, the optimal value of the regularization parameter α is critical: if too small, the solution x_α will be oscillatory; if, on the contrary, too large, the solution will be over-smoothed.

There is another viewpoint to consider problem (10.14), i.e. in terms of a two-objective minimisation. In fact, the norm $\|Ax_\alpha - b\|_2^2$ in the Tikhonov functional (10.13) accounts for the agreement of the field model to the supplied data. When the norm itself is minimised, the agreement becomes very good, but the solution is likely to be unstable. That is where the second norm $\|x_\alpha\|_2^2$ appearing in (10.13) comes in, in order to control the smoothness of the solution, i.e. its stability with respect to perturbations in the data. In turn, minimising the second norm by itself gives a very smooth solution that might have nothing in common with the given data. Therefore, the trade-off curve of the best compromises between agreement and smoothness is to be sought for, by varying the regularization parameter α in a suitable way and then selecting an equilibrium point along the curve. As a results, a typically L-shaped curve is obtained: various techniques of regularization are based just on the use of the L-curve [6].

10.8 Optimal Synthesis of Current Distribution in a Small Solenoid

Suitably arranged current-carrying coils can be used to synthesize a magnetic field with the required distribution. In biomedical engineering there are several applications: for instance, a homogenous magnetic field is the background of nuclear magnetic-resonance spectroscopy. Moreover, in magneto-fluid hyperthermia (MFH) the field homogeneity helps the uniform distribution of heat generated in the nano-particle fluid previously injected in the target region, like e.g. a tumor mass under treatment [1]. A case study of current synthesis in a small solenoid by means of Tikhonov regularization is here presented.

Let a single-layer solenoid of finite length be considered. A non-trivial inverse problem is the synthesis of the magnetic field along the solenoid axis. This problem can be formulated as follows: find the current density $J(y)$, $-d \le y \le d$ that originates the prescribed distribution $B_0(y)$ of the induction field in a one-dimensional sub-region $-c < y < c$ on the solenoid axis. It is assumed that the solenoid is supplied by a winding, sectioned in a number of equal coils along the y axis, in such a way as to allow different currents in different coils. After integrating the equation which gives the induction field due to an elementary ring with internal radius r_i, external radius r_s, and carrying current $J(\xi)d\xi$, the following expression for the induction field at point y due to the multi-coil winding holds

$$B(y) = \frac{\mu_0}{2}\int_{-d}^{d}\int_{r_i}^{r_s}\frac{J(\xi)r^2 dr d\xi}{\sqrt{\left[r^2 + (y-\xi)^2\right]^3}} \tag{10.18}$$

If $B(y) = B_0(y)$, $-c \leq y \leq c$ is the prescribed function and $J(\xi)$, $-d \leq \xi \leq d$ is the unknown function, then (10.18) proves that the field distribution which is required on the solenoid axis can be synthesized by means of a suitable current distribution in the winding coils. Specifically, a small-size solenoid, useful for in vitro experiments of MFH, is here considered: $d = 15$ mm, $r_i = 9$ mm, $r_s = 10$ mm, $c = 5$ mm. The winding is composed of 20 coils: therefore, 20 unknown currents (design variables) are to be identified. In turn, the induction field is prescribed in 41 evenly spaced points along the solenoid axis. Due to the linear relationship between current distribution J and induction field B in (10.18), the synthesis problem can be cast as the solution of a rectangular system of equations

$$Ax = b \tag{10.19}$$

where A is the (41,20) matrix linking the field at the probe points with the current in the turns, x the vector of unknown currents, while b is the vector of prescribed induction values at the probe points: put simply, (10.19) is the discretized version of (10.18). In particular, a uniform field distribution $B_0 = 0.84$ mT is prescribed. The problem is an ill-posed one because the rank of (20,20) normal matrix A^tA is equal to $11 < 20$.

In Fig. 10.1 the curve relating the residual norm with the quasi-solution norm as a function of regularizing parameter α is shown: the marked point corresponds to the optimal value of α, i.e. the lowest value of α such that augmented matrix $A^tA + \alpha I$ is a full-rank one. It could be interpreted as the best trade-off between two conflicting criteria, namely: the residual norm, which controls accuracy, and the quasi-solution norm, which controls instability against small perturbation of data.

In Fig. 10.2 the corresponding distribution of current per turn is shown.

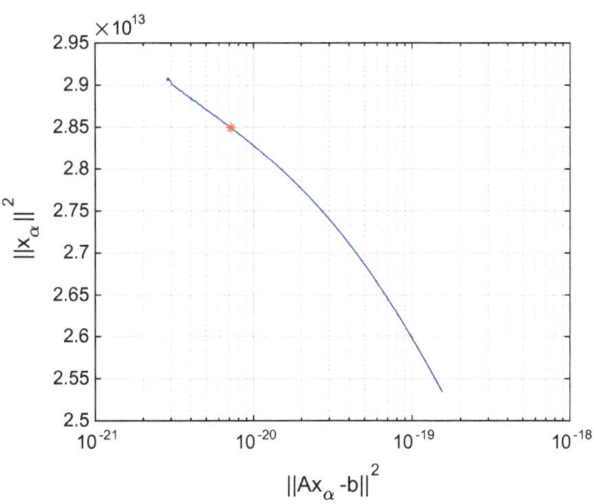

Fig. 10.1 Curve relating the residual norm with the quasi-solution norm as a function of regularizing parameter α: the marked point corresponds to the optimal value $\alpha_{opt} = 7.8 \times 10^{-33}$

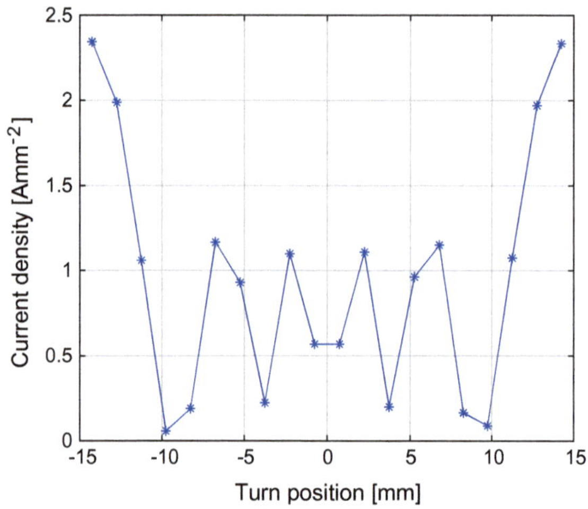

Fig. 10.2 Current distribution originating uniform field distribution B_0 along the solenoid axis when the regularizing parameter value is $\alpha = \alpha_{opt}$

References

1. Bertani R, Ceretta F, Di Barba P, Forzan M, Dughiero F, Michelin R, Sgarbossa P, Sieni E, Spizzo F (2015) Optimal inductor design for nanofluid heating characterisation. Eng Comput 32(7):1870–1892
2. Di Barba P (2010) Multiobjective shape design in electricity and magnetism. Springer
3. Di Barba P, Savini A, Wiak S (2008) Field models in electricity and magnetism. Springer, Berlin, Germany
4. Groetsch CW (1999) Inverse problems. The Mathematical Association of America
5. Hadamard J (1923) Lectures on the cauchy's problem in linear partial differential equations. Yale University Press
6. Hansen PC, O'Leary DP (1993) The use of L-curve in the regularization of discrete ill-posed problems. SIAM J Sci Comput 14:1487–1503
7. Neittaanmäki P, Rudnicki M, Savini A (1996) Inverse problems and optimal design in electricity and magnetism. Oxford science Publications
8. Tikhonov A, Arsénine V (1976) Méthodes de Résolution de Problèmes Mal Posés, MIR Editions

Chapter 11
Numerical Methods for MEMS Design: Automated Optimization

Paolo Di Barba and Maria Evelina Mognaschi

11.1 Solving Inverse Problems by Means of Numerical Optimization: Single-objective Formulation

As stated in Sect. 10.3, the problem of identifying or reconstructing a given quantity, based on known data e.g. measurements, is called an inverse problem. Loosely speaking, an inverse problem is one in which an effect is measured and the cause of it is to be determined. Inverse problems for which data come from measurements are known as identification problems.

Another group of inverse problems refers to synthesis problems in which data are assumed arbitrarily. Optimal design problems, where the purpose is to design a device which can provide a defined behaviour or optimal performance, belong to this group. In particular, when the geometry of a device has to be optimized, a shape design problem, also called "shape synthesis", occurs. In Chap. 14 shape design problems for MEMS devices are solved by means of methods shown and discussed in this Chapter.

The n unknowns x of an inverse problem are normally called design variables. They are real values, although in some cases they are integer, belonging to a feasible region Ω. In multivariate problems, $n > 1$.

The design variables may be geometric coordinates of the field region for shape design problems, because the design variables parametrize the geometry [7]; in turn they could be values of sources [11] or parameters characterizing the region (for identification problems) or whatever.

The solution of the inverse problem is generally performed by means of the optimization (minimization) of a suitable function *f(x)* called objective function or cost function or design criterion:

$$\text{given}\, x_0 \in \Omega \quad \text{find} \inf_x f(x), \quad x \in \Omega \tag{11.1}$$

where x_0 is the initial guess; properly speaking, (11.1) is a problem of unconstrained optimization. Normally, it is assumed that f is bounded in Ω.

P. Di Barba and S. Wiak, *MEMS: Field Models and Optimal Design*, Lecture Notes in Electrical Engineering 573, https://doi.org/10.1007/978-3-030-21496-8_11

Function f may represent some performance depending on the field or simply the discrepancy between computed and known field values (error functional), that is the residual $\|Ax - b\|$ in Sect. 10.5.

When it is required to maximize an objective function, it must be considered that

$$\sup_x f(x) = \inf_x[-f(x)] \tag{11.2}$$

In general, the objective function f, which depends on the field, is not known analytically. Consequently, the classical conditions of optimality (i.e. null gradient and positive-definite Hessian matrix) cannot be applied a priori, because the objective function is known only numerically as a set of values at sample points. Moreover, in general, f is neither convex nor differentiable or smooth. Therefore, it is not guaranteed to get solutions; in particular, f might exhibit some local minima in addition to the global one. Any way, a solution to (11.1) can be obtained just numerically and the procedure may be troublesome and time-consuming.

As said before, the numerical solution of inverse problems in electricity and magnetism require, as a rule, a routine for calculating the field, which is integrated with a routine minimizing the objective function [15, 17].

Usually, the device or system to be optimized is represented by a finite-element model in two or three dimensions. The main flow of the computation is driven by the minimization routine, which in the simplest way is carried out step by step. Starting from x_0, an iterative procedure updates the current design point x_k as

$$x_{k+1} = x_k + \lambda s_k \tag{11.3}$$

where λ is a scalar and s_k is the current search direction within the feasible region. Given x_{k+1}, the routine of field analysis generates a new finite-element grid, the field simulation is restarted and the evaluation of $f(x)$ is so updated.

At the end of computation, the result could represent either a local minimum or simply a point which is better than the initial one because f has decreased; in the latter case, a mere improvement (and not the optimization) of f has been achieved. In general, the optimization trajectory may converge to different local minima, depending on the initial point, and the global optimum cannot be deduced from the local behaviour of the objective function.

In a more advanced formulation, the objective function should fulfill constraints, which may be expressed as inequalities, equalities and bilateral bounds. Formally, the problem can be stated as follows:

$$\text{given } x_0 \in \Omega \quad \text{find} \inf_x f(x), \quad x \in \Omega \tag{11.4}$$

subject to

$$g_j(x) \le 0 \; j = 1, \ldots, n_i$$
$$h_j(x) = 0 \; j = n_i + 1, \ldots, n_c$$

$$\ell_k \leq x \leq u_k \; k = 1, \ldots, n_b \tag{11.5}$$

Constraints and bounds set the limits of the feasible region Ω. A simple technique to handle constraints is to transform the constrained problem into an unconstrained one, by adding a penalty term to the objective function, when the design variables violate the constraints. This way, a sequence of unconstrained problems is solved, which is assumed to converge to the solution of the constrained problem.

A cost-effective and accurate solution to the optimization problem depends on the number of design variables and constraints, as well as on the properties of objective function and constraints.

11.2 Multi-objective Formulation: Pareto Optimality

In some cases multiple objective functions are prescribed simultaneously. Problems of this kind belong to the category of multi-objective or multi-criteria optimization. Several design problems in electricity and magnetism are characterized by a vector of n_f objective functions in mutual conflict, for which the most general solution is represented by the Pareto front of non-dominated solutions, i.e. those for which the decrease of a function is not possible without the simultaneous increase of at least one of the other functions. Non-dominated solutions are called also non-inferior or efficient solutions.

Formally, considering n_v variables, a multiobjective optimization problem can be cast as follows:

$$\text{given}\, x_0 \in \Omega \quad \text{find} \inf_x F(x), \quad x \in \Omega \tag{11.6}$$

subject to (11.5).

In (11.6), $F(x) = \{f_1(x), \ldots, f_{n_f}(x)\} \subset R^{n_f}$ is the objective vector, assuming $n_f \geq 2$. Therefore, F defines a transformation from the design space R^{n_v} to the corresponding objective space R^{n_f}. It is assumed that the objectives are bounded and conflicting, namely

$$\exists x_i^* \text{ such that } \; f_i(x_i^*) = \inf f_i(x), \quad i = 1, \ldots, n_f \tag{11.7}$$

and $x_i^* \neq x_j^*, \quad i \neq j, \quad j = 1, \ldots, n_f$.

In general, the utopia solution

$$U = \{U_i\} = \{\inf f_i(x)\}, \quad i = 1, \ldots, n_f \tag{11.8}$$

minimizing all f_i does not exist and non-dominated solutions $\tilde{x} \in \Omega$ are accepted. It can be noted that the Utopia point U is out of the feasible design region Y (Fig. 11.1a). In a symmetrical way, the Anti-Utopia point A (see Fig. 11.1a) can be defined as the

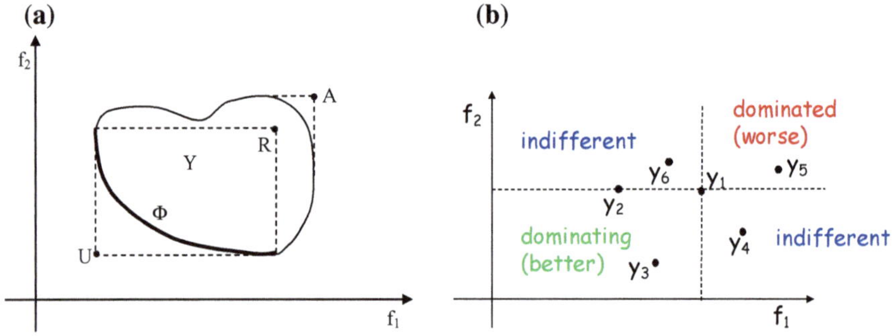

Fig. 11.1 Objective space (**a**) and Pareto dominance criterion (**b**)

solution which maximizes all f_i, while the Nadir point R is the mirror image of the Utopia point [3].

In this respect, given two feasible solutions $x_a \in \Omega$ and $x_b \in \Omega$, x_a is said to dominate x_b if $f_i(x_a) \leq f_i(x_b), \quad i = 1, \ldots, n_f$ and $f_i(x_a) < f_i(x_b)$ for at least a value of index $i = 1, \ldots, n_f$. Now, let $P \subset \Omega$ be a set of non-dominated solutions $\tilde{x}_a$; if, for any $\tilde{x}_a \in P$, there is no $x_b \in \Omega$ dominating $\tilde{x}_a$, then P represents the Pareto set and the corresponding image $\Phi = F(P)$ is the Pareto front; the latter is called also trade-off curve.

For the sake of an example, in Fig. 11.1b six solution points and the dominance dihedral related to solution y_1 are shown in a two-dimensional space: y_3, which belongs to the dominance dihedral, dominates y_1, while y_2, y_4 and y_6 are indifferent with respect to y_1 and, finally, y_5 is dominated by y_1.

Traditionally, the multiobjective problem is reduced to a single-objective one by introducing a preference function $\psi(x)$, e.g. the weighted sum of $f_i(x)$:

$$\psi(x) = \sum_{i=1}^{n_f} c_i f_i(x), \quad 0 < c_i < 1, \quad \sum_{i=1}^{n_f} c_i = 1 \tag{11.9}$$

However, in the last decades, attention is paid to methods able to solve a multi-objective optimization problem considering separately the two objective functions and, recently, new methods for solving the so-called many-objective optimization problems (more than two or three objective functions) are under study [10, 15].

Accordingly, in this book, all the proposed multi-objective problems are solved in terms of identifying the relevant Pareto front.

11.3 Gradient-Free and Gradient-Based Methods

Several algorithms for both unconstrained and constrained optimization are available; basically, they can be sorted into two broad classes, i.e. gradient-free and gradient-based methods; in the former case, the information about the derivative is not used while in the latter case it is used.

Methods that use only function evaluations (zero-order methods) fall under the first class; they are suitable for problems characterized by non-linearity and discontinuities of the objective function. The computational efficiency is low, due to repeated calls to the objective function.

In turn, first-order gradient-based methods basically follow the direction of the steepest descent; they are more efficient than the zero-order ones. when regular objective functions are dealt with. The simplest way to approximate the gradient of the objective function relies on finite differences; e.g. if $g_i(x_i) \equiv \frac{\partial f}{\partial x_i}$ is the i-th component of the gradient, a forward difference gives

$$g_i(x_i) \cong \frac{f(x_i + h) - f(x_i)}{h}, \quad i = 1, \dots, n \tag{11.10}$$

where h is the incremental step. However, it should be noted that the approximation of the gradient is expensive and represents also an additional source of numerical ill conditioning (round-off errors in the computation of both objective functions and finite differences).

Higher-order methods, like Newton's method, are rarely used because they are suitable only when the Hessian matrix can be easily computed.

Regardless of the order, all the aforementioned methods are local in a sense, because they are able to identify the closest minimum to the starting point, which is a local one, unless f is convex. For this reason they are said to perform a deterministic search.

To cope with these difficulties, non-deterministic minimization algorithms, which are derivative-free and perform a stochastic search, have been developed.

Non-deterministic methods offer remarkable advantages over methods that use only local information to improve the current solution. In fact, they are robust, reliable and suitable for non-convex, non-smooth and discontinuous functions, also with discrete-valued variables. In particular, they give a chance to approximate the global minimum of the objective function, regardless of the starting point. Another advantage is that they exhibit an inherent parallelism. The drawbacks are the huge computational effort and the slow convergence.

Since they have an heuristic background, it has to be pointed out that for non-deterministic methods convergence is proven just in numerical terms and not on theoretical basis, contrary to what happens for their deterministic counterpart.

In the non-deterministic category, the most popular methods are: simulated annealing, evolution strategies, genetic algorithms. More recently, a class of nature-inspired methods has been developed, among which: ant colony, particle swarm, artificial immune systems.

An example of gradient-free method is the simplex method. It is based on the comparison among the cost function values at the $n + 1$ vertices of a polytope (simplex), where n is equal to the dimension of the search space. In the case of $n = 2$ ($n = 3$), the polytope is a triangle (a tetrahedron).

The algorithm begins with $n + 1$ points, which form the starting polytope, and the calculation of the associated objective function values. At each iteration a new polytope is set up by generating a new point to replace the worst vertex of the old polytope, i.e. the vertex corresponding to the highest value of objective function. Specifically, the worst vertex is replaced by its reflection with respect to the remaining n vertices. If the objective function evaluated at the new point is higher than at the worst vertex, then the new point is rejected and the vertex with the second worst value is reflected.

When it happens that a vertex belongs to the polytope for a number of iterations which exceeds the given one, then the polytope is updated by contraction. The whole procedure is iterated until the diameter of the simplex is less than the specified tolerance.

11.4 Evolutionary Computing: The Genetic Paradigm

Evolution strategy mimics the survival of the fittest individual that is observed in nature. An algorithm of the lowest order (i.e. a single parent generates a single offspring) is here shortly presented. The search in the design space begins in a region of radius d_0 (standard deviation) centered at the initial point m_0 (mean value); m_0 is externally provided, while d_0 is internally calculated on the basis of the bounds boxing the variation of the design variables.

Setting $m = m_0$ and $d = d_0$ at the initial iteration, the *generation* of the design vector $x = m + ud$ then proceeds, resorting to a stochastic sample $u \in (0, 1)$; generally, u is a normally distributed sample. It is verified that x fulfils bounds and constraints (i.e. that x is feasible), otherwise a new design vector is generated until it falls inside the feasible region.

The associated objective function $f(x)$ is then evaluated and the test $f(x) < f(m)$ is performed; if the test is successful, m is replaced by x (the so-called *mutation* or *selection* process), otherwise m is retained.

The next step is concerned with the size of the search region that will be used for the successive iteration. The underlying rationale is that when a point better than the current one is found, the radius of the search region is increased around the new point to search for further improvements; if no improvement is found, the radius of the search region is gradually decreased up to convergence (*annealing* process).

In this respect, the evolutionary algorithm substantially differs from a deterministic one, in which the search region would be narrowed around the better point in order to converge towards the corresponding, nearest minimum. The drawback, in the latter case, is that this minimum might be a local one. On the contrary, the evolutionary algorithm, if successful in finding a better point, covers a larger region of search in order to see if there might be another good candidate in the neighborhood, and then does the opposite when this is not possible. This way, there is a non-zero probability of finding the region where the global optimum of the objective function is located.

The *annealing* process is ruled by the history of the k previous iterations, used to establish a trend: if at least a fraction p of the last k iterations were successful (an iteration is successful if x is feasible and improves the objective function), then the trend is said to be positive, while it is negative otherwise. If the trend is positive, the radius d of the search region is set to $q^{-1}d$; otherwise, it is set to qd. In particular, during the first k iterations, d remains unchanged.

The procedure stops when the prescribed accuracy $\left|\frac{d}{d_0}\right|$ is achieved. Quantities p and q are named probability of success and rate of annealing, respectively and represent the "tuning knobs" of the algorithm; heuristic values for k, p and q arc 50, 0.2 and 0.8–0.9, respectively.

11.4.1 Multi Objective and Pareto-like Evolution Strategy

It is based on an extension of the zero-order evolution strategy, where the mutation operator is modified according to the concept Paretian optimality. Specifically, an offspring solution is accepted if and only if it strongly dominates the parent solution for all the objectives, provided problem constraints are fulfilled. As a consequence, the algorithm finds a trajectory connecting the start solution to a non-dominated solution, which is assumed to approximate a Pareto-optimal solution in the n_f -dimensional objective space ($n_f = 2$ for MOESTRA [8] and $n_f \geq 2$ for P-Estra), [14]. The algorithm stops when the search radius in the design space is smaller than the prescribed tolerance.

11.5 Nature-Inspired Computing: Wind-Driven Optimization

Wind Driven Optimization (WDO) is inspired by the atmospheric motion of air parcels [1]. In fact, wind blows in a way to equalise imbalances in the air pressure; likewise, in optimisation the design points are moved from high-gradient to low-gradient positions. Accordingly, a swarm of $p > 1$ artificial air parcels, which are randomly distributed over the n-dimensional design space and assigned random velocities, is considered. The WDO algorithm is governed by the following equation:

$$u_i(t_{k+1}) = (1-\alpha)u_i(t_k) - gx_i + \beta\left|1 - i^{-1}\right|\left(x_{\mathrm{opt}} - x_i\right) + 2u^{\perp} \tag{11.11}$$

where the left-hand side gives the velocity of the i-th out of p air parcels at time t_{k+1}. The right-hand side is characterised by the following four terms: the inertial term, depending on velocity at time t_k and frictional coefficient α (conservative operator); the gravitational-like term g_x, which biases the current position x_i towards the gravity centre of the design space (pull-in operator); the pressure-gradient term (the main operator), where index $i \geq 1$ is proportional to the pressure value at position x_i while x_{opt} is the position of the lowest pressure found in the previous $k - 1$ iterations; the Coriolis-like acceleration term, due to which the velocity of the i-th parcel is influenced by the orthogonal velocity of another, randomly selected parcel (information-exchange operator). Positive-valued constants (α, β, g) are algorithm-dependent parameters. The position of the ith air parcel at iteration $k + 1$ is then updated as

$$x_i(t_{k+1}) = x_i(t_k) + u_i(t_{k+1}) \tag{11.12}$$

and the boundaries are checked to prevent any air parcel from exiting the design space. The procedure continues until the maximum number of iterations is reached. Finally, the best pressure location at the end of the last iteration is recorded as the optimisation result. The WDO is able to approximate the solution to a single-objective optimisation problem, provided pressure value and air parcel position correspond to objective function and design variable vector, respectively. To generalise the algorithm, and make it able to solve a multi-objective optimisation problem exhibiting $m > 1$ objective functions in conflict (M-WDO), the following modifications are needed. The most general solution is given by the Pareto front of non-dominated solutions, i.e. the set of solutions such that no decrease of a function is possible without the simultaneous increase of at least one of the other $m - 1$ function. To this end, in M-WDO a generalised pressure, which takes into account simultaneously m objective functions, is defined [4, 9]. At the current iteration, air parcels in the swarm are mapped from the n-dimensional design space X to the m-dimensional objective space F, where coordinates are just the objective function values ($f_1(x)$, ..., $f_m(x)$) of design vector (x_1, ..., x_n). In F space, air parcels are sorted according to the following criterion. The air parcels in the swarm corresponding to non-dominated solutions are assumed to identify the first sub-front, which is assigned rank index equal to one. Then non-dominated solutions are temporarily ignored to process the rest of air parcels in the swarm. The procedure is continued until all air parcels are sorted in sub-fronts. By definition, all solutions within a sub-front are equivalent and, therefore, they are assigned the same generalised-pressure value, equal just to the rank index of the sub-front they belong to. This way, in the subsequent iteration the algorithm will both improve the solutions of the first sub-front and move dominated solutions towards the first sub-front, in the attempt to minimise the generalised pressure of each air parcel in the swarm. So, the governing equation is basically (11.11); however, factor $|1 - i^{-1}|$ must be updated as $|1 - j^{-1}|$, where $j \geq 1$ is the rank index of the sub-front the ith solution belongs to. A simplified pseudo-code follows:

(0) start;
(1) initialise $p > 1$ air parcels and define $m > 1$ objective functions;
(2) set design space boundaries and max number of iterations;
(3) assign random position and velocity;
(4) compute objective functions and map air parcels in F space;
(5) sort air parcels in non-dominated sub-fronts;

(6) evaluate generalised pressure for each air parcel;
(7) update velocity;
(8) update position and check design space boundaries;
(9) save the non-dominated solutions; If the max number of iterations is reached, then go to step 10; else go to step 4 and continue;
(10) end.

The whole procedure gives rise to a quick convergence towards the nondominated region of the objective space. Moreover, the computational cost of M-WDO algorithm is the same as WDO, i.e. proportional to the number of air particles times the maximum number of iterations. Most of runtime is due to the computations of objective functions, which generally require to solve a field analysis problem.

11.6 Biogeography-Based Optimization

Computational biogeography models the process of natural immigration and emigration (or extinction) of species between small islands in the search for more friendly habitats, which is observed in nature [18]. The assumption of small island is important in order to exclude surface-related local perturbations, like e.g. climate variations, that might make a subregion most preferred to other subregions of the same island.

Each solution considered in a biogeography-inspired optimization algorithm (BBO) is treated as a habitat (design vector) composed of suitability index variables (SIV, design variables), and each habitat exhibits a quality given by the habitat suitability index (HSI, objective function) [21]. Remarkably, in contrast to GA based algorithms, the original population is not discarded after each generation, but it is progressively modified by means of two stochastic operators, *i.e.* migration and mutation: migration improves the HSI of poor habitats by sharing features from good habitats; in turn, mutation modifies some randomly selected SIV of a few habitats in view of a better exploration of the ecosystem (design space). At each generation BBO exploits the HSI of each habitat based on its migration rate, while the emigration rate is set to be complementary to immigration. This way, the HSI of each habitat is improved (see Fig. 11.2).

In particular, at each iteration, habitats are sorted from the best ones to the worst ones according to the relevant value of the fitness. For each SIV of each habitat a random event r_j, such that $0 < r_j < \sup(I,E)$ with I maximum immigration rate and E maximum emigration rate, is generated. Then value r_i is compared with the corresponding immigration rate λ_i of the considered habitat. If $r_j > \lambda_i$ then immigration occurs: the SIV considered in the current ecosystem migrates to the next ecosystem, keeping the same location (Fig. 11.3a).

In contrast, if $r_j < \lambda_i$ emigration occurs (Fig. 11.3b): the current SIV in the considered habitat (say the ith habitat) goes extinct and the SIV of another habitat (say the kth habitat) takes the same SIV position in the ith habitat of the next ecosystem. The kth habitat is selected depending on the emigration rate and the SIV.

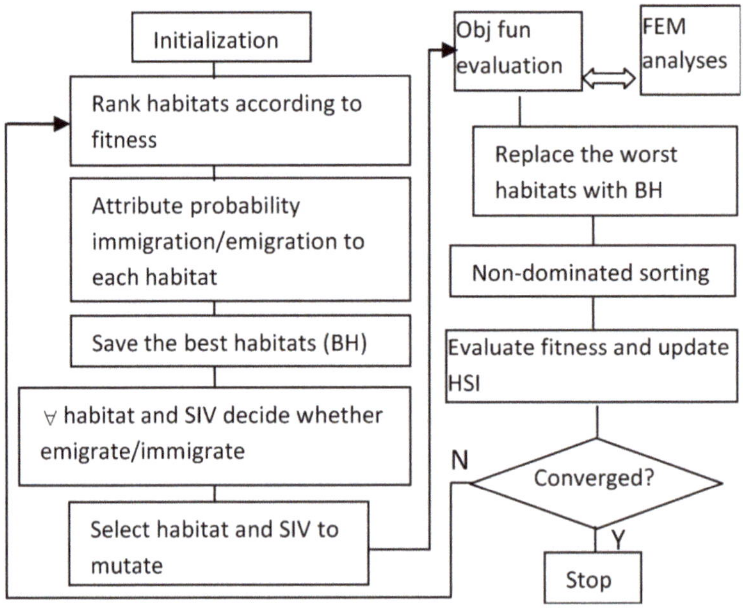

Fig. 11.2 Flow chart of the proposed BiMO algorithm

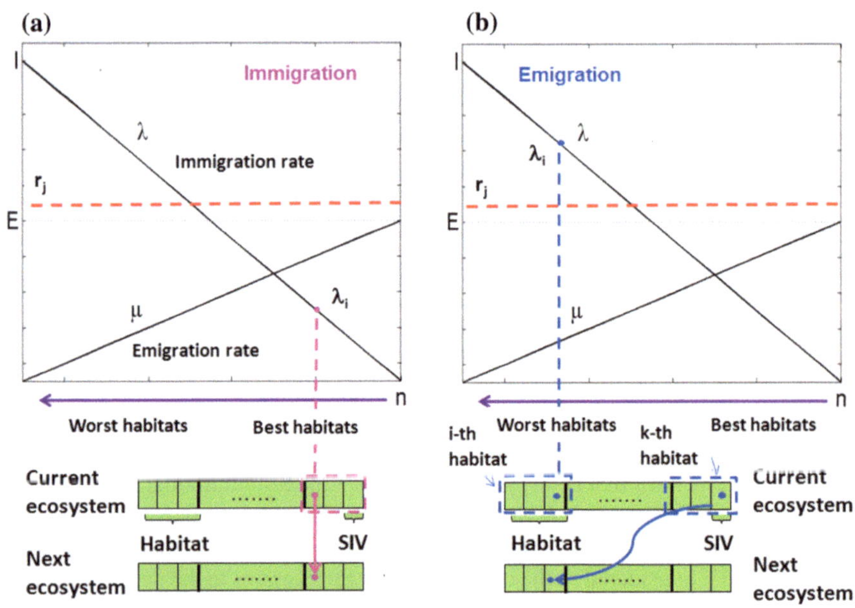

Fig. 11.3 Schemes of immigration and emigration events

In computational electromagnetics, applications and also modifications of BBO are an emerging new field of research: various single-objective problems of optimal design have been successfully solved; moreover, attempts to transform BBO in a multi-objective optimization algorithm have been considered [20, 22, 23]. In particular, the concept of predator-prey has been implemented for hunting the worst individuals per objective function and making the surviving individuals spread in the objective space in order to escape from the predator [2]. Though being interesting and effective, similar approaches might suffer from some limitations, like *e.g.* the possible loss of some non-dominated solutions (the worst individual per objective function might well be the end of the current set of non-dominated solutions), and the lack of an explicit criterion ruling the spread or crowd of the surviving individuals in the objective space.

As an alternative, in this Chapter, a modification of the definition of HSI by means of a generalized fitness is shown [5, 6, 12, 13, 16]. It takes into account simultaneously two or more objective functions by exploiting the concept of non-dominated ranking of solutions in the objective space. At the current iteration, habitats are ordered according to two criteria: first, they are subdivided in locally non-dominated sub-fronts which in turn are ranked from the first to the last. Next, habitats are ranked within the sub-front they belong to, taking into account their crowding distance; in fact, crowded habitats are less preferred than habitats which are regularly spaced.

In general, let a population of $n_p > 1$ individuals, each individual provided with $n_f > 1$ objective function values, be considered; a non-dominated set of solutions is found based on the following algorithm:

0) initialize k=1;
i) begin with i=1;
ii) for j=1,n_p and $i \neq j$, compare individuals x_i and x_j according to the definition of dominance for all n_f objectives;
iii) if, for any j, x_i is dominated by x_j , consider x_i as "dominated";
iv) if all individuals have been considered, go to step v;
else set i as i+1 and go to step ii;
v) all individuals that are not marked as "dominated" are non-dominated solutions belonging to front k;
vi) remove non-dominated solutions, set k=k+1 and go to i);
else end .

All the individuals found at step v are assumed to identify the k-th non-dominated front in the population and are assigned a generalized fitness value (e.g. equal to k, the index of the current front). At a first glance, the same fitness value is assigned to all non-dominated individuals in the k-th front, to give them an equal reproductive potential. In order to maintain diversity in the population, the non-dominated solutions are then shared with their fitness values. Basically, sharing is achieved by dividing the fitness values by a quantity (called niche radius) inversely proportional to the number of individuals around it. This procedure causes multiple optimal points to co-exist in the population. Once they have been identified, a fitness value, which is higher than k and smaller than $k + 1$, is assigned to each individual. Then, shared fitness values are updated as before.

After sharing, non-dominated individuals are temporarily ignored to process the rest of population. This step-by-step procedure is iterated to find the $(k + 1)$th front of non-dominated individuals in the population. The process is continued until all population individuals are assigned a shared fitness value.

Finally, elitism is exploited by preserving a part of the non-dominated habitats at the current generation.

The μBiMO method, an extension of the biogeography-inspired multi-objective optimisation (BiMO) algorithm, was proposed recently [19]. It is suitable for solving multi-objective optimisations in the field of industrial design, where the objective functions are evaluated by means of numerical methods as the FEM. In fact, it is based on a small number of islands (hence the name μBiMO), i.e. few objective function calls are required. The algorithm so modified leads to a strong reduction of computational time, and, at the same time, improves the quality of results.

References

1. Bayraktar Z, Komurcu M, Bossard JA, Werner DH (2013) The wind driven optimization technique and its application in electromagnetics. IEEE Trans Antennas Propag 61(5):2745–2757
2. Costa Silva MA, Coelho LS, Lebensztajn L (2012) Multiobjective biogeography-based optimization based on predator-prey approach. IEEE Trans Magnetics 48(2):951–954
3. Di Barba P (2010) Multiobjective shape design in electricity and magnetism. Springer
4. Di Barba P (2016) Multi-objective wind-driven optimisation and magnet design. Electron Lett 52(14):1216–1218
5. Di Barba P, Dughiero F, Forzan M, Mognaschi ME, Sieni E (2018) New solutions to a multi-objective benchmark problem of induction heating: an application of computational biogeography and evolutionary algorithms. Arch Electr Eng 67(1):139–149
6. Di Barba P, Dughiero F, Mognaschi ME, Savini A, Wiak S (2016) Biogeography-inspired multiobjective optimization and MEMS design. IEEE Trans Magn 52(3)
7. Di Barba P, Gotszalk T, Majstrzyk W, Mognaschi ME, Orłowska K, Wiak S, Sierakowski A (2018) Optimal design of electromagnetically actuated MEMS cantilevers. Sensors (Switzerland) 18(8)
8. Di Barba P, Mognaschi ME (2009) Industrial design with multiple criteria: shape optimization of a permanent-magnet generator. IEEE Trans Magn 45(3):1482–1485
9. Di Barba P, Mognaschi ME, Przybylski M, Rezaei N, Slusarek B, Wiak S (2018) Geometry optimization for a class of switched-reluctance motors: a bi-objective approach. Int J Appl Electromagnet Mech 56(S1):S107–S122
10. Di Barba P, Mognaschi ME, Rezaei N, Lowther DA, Rahman T (2019) Many-objective Shape Optimisation of IPM Motors for Electric Vehicle Traction. in press on Int J Appl Electromagnetics Mech IJAEM
11. Di Barba P, Mognaschi ME, Savini A (2007) Synthesizing a field source for magnetic stimulation of peripheral nerves. IEEE Trans Magn 43(11):4023–4029
12. Di Barba P, Mognaschi ME, Savini A, Wiak S (2016) Island biogeography as a paradigm for MEMS optimal design. Int J Appl Electromagnetics Mech IJAEM 51(s1):97–105
13. Di Barba P, Mognaschi ME, Venini P, Wiak S (2017) Biogeography-inspired multiobjective optimization for helping MEMS synthesis. Arch Electr Eng 66(3):607–623

14. Di Barba P, Mognaschi ME, Wiak S, Przybylski M, Slusarek B (2018) Optimization and measurements of switched reluctance motors exploiting soft magnetic composite. Int J Appl Electromagnet Mech 57(S1):S83–S93
15. Di Barba P, Savini A, Wiak S (2008) Field models in electricity and magnetism. Springer, Berlin, Germany
16. Di Barba P, Savini A, Wiak S (2017) Higher-order multiobjective design of MEMS. Int J Appl Electromagnet Mech 53(S2):S239–S247
17. Di Barba P, Wiak S (2015) Evolutionary computing and optimal design of MEMS. IEEE/ASME Trans Mechatron 20(4):1660–1667
18. Macarthur RH, Wilson EO (1967) The theory of island biogeography. Princeton University Press
19. Mognaschi ME (2017) Micro biogeography-inspired multi-objective optimisation for industrial electromagnetic design. Electron Lett 53(22)
20. Roy PK, Ghoshal SP, Thakur S (2010) Multi-objective optimal power flow using biogeography based optimization. Electric Power Compon Syst 38:1406–1426
21. Simon D (2008) Biogeography-based optimization. IEEE Trans Evol Comput 12(6):702–713
22. Singh U, Kumar H, Kamal TS (2010) Design of Yagi-Uda antenna using biogeography based optimization. IEEE Trans Antennas Propag 58(10):3375–3379
23. Singh S, Mittal E, Sachdeva G (2012) NSBBO for gain-impedance optimization of Yagi-Uda antenna design. Proc World Congress Inf Commun Technol 2012:856–860

Chapter 12
From MEMS to NEMS

Teodor Gotszalk

12.1 Introduction

Nanotechnology, as the scientific and technological discipline dealing with the design, fabrication and application of systems whose dimensions or tolerances are in the domain of nanometers, is becoming increasingly important in many industrial and scientific areas. Nanotechnologies and nanoscience are triggered by diverse fields and applications but on the other hand, they trigger by themselves future industrial and practical solutions. One of the most important challenges observed nowadays in nanotechnology is driving the manufacturing processes to sub-nm accuracy level for critical features and positioning tasks.

The progress in nanotechnology is directly connected with the progress in the Micro-Electro-Mechanical Systems (MEMS) technology, which in many countries is also called as the Microsystems technology. In general it can be described as the technology of miniaturized electro-mechanical devices and structures that are fabricated using technologies used for fabrication of semiconductor integrated circuits (ICs). To the MEMS belong simple devices like supported beams (cantilevers), double-clamped microbridges and membranes as well as complicated structures with moving mechanical components, whose deflection is precisely detected and controlled. In the most mature form the electromechanical components are integrated with microcontrollers, which perform not only simple operations but enable implementation of artificial intelligence (AI) algorithms.

The critical physical dimensions of the MEMS can vary from several microns to several millimeters. The attractiveness of the MEMS technology is intuitive—the small dimensions of the MEMS device make the MEMS device sensitive to phenomena occurring at micro- and nanoscale. In other words the smaller structure one wishes to observe the smaller tool must be used. Moreover, as the MEMS are fabricated in the so called batch processes, their cost can be reduced and the scale of integration increased, leading to the increase of the observation throughput. It should be noted however, that the MEMS are distinct from the idea of the molecular electronics and molecular nanotechnology. In principle, the MEMS are fabricated in the so called

P. Di Barba and S. Wiak, *MEMS: Field Models and Optimal Design*, Lecture Notes in Electrical Engineering 573, https://doi.org/10.1007/978-3-030-21496-8_12

top-down processes, in which a device is assembled by various processes shaping the final mechanical and electrical device form. In contrast, bottom-up processes, are based on the chemical properties of individual species, which cause single-molecule components to self-organize and/or self-assemble into some useful conformation. It should be noted however, that the MEMS make it possible to manipulate with single molecules and from that point of view the MEMS are of huge importance for the bottom-up technology.

In the natural way the MEMS technology merges at the nanoscale into Nano-Electro-Mechanical-Systems (NEMS). The experimental possibilities are in this case even bigger as the tool dimension size is reduced [15]. Challenge of the today's MEMS/NEMS technology is that it does not only involve application of modern microfabrication techniques but also require thorough analysis of the system operation. The analysis based on classical physics cannot often be used to describe and interpret the phenomena defining the functionality of a MEMS/NEMS device. At these scales of dimensions the large surface area to volume ratio of the MEMS devices, surface effects such as electrostatics, wetting, molecular adhesion, which includes chemical interactions, dominate volume effects such as inertia or thermal mass. Due to very small device dimensions and tiny distances between the device components, which are being moved, quantum phenomena must be taken into account in order to describe the recorded phenomena. This in turn makes the interpretation of the observed MEMS/NEMS behavior distinct from the interpretation the engineers are today accustomed to. In Table 12.1 a group of MEMS/NEMS devices is presented illustrating the described tendency. The list begins with a MEMS cantilever whose actuation is driven electromagnetically [26]. The spring beam is formed out of crystalline silicon, a metal thin film loop is used to actuate the structure deflection. The simplest way to fabricate the structures whose thickness is of hundreds of nanometers is to form them out of a thin film. In this way the mass of the structure can be reduced as in case of the silicon nitride microbridges [23]. Silicon nanowires are one the next NEMS device examples [38]. When released out of the substrate they can act as nanoresonators. The most important feature of the proposed architecture is that each of the wires integrated in an array can be addressed separately, which increases the operation throughput, as the response of every wire can be sensed separately.

Besides classical materials utilized and applied in micro- and nanoelectronics, there are also attempts to apply the 2 dimensional (2D) crystals as the materials for the MEMS and NEMS devices. In such a setup, the atomically thick membranes form the NEMS mechanical part. Excellent material properties of 2D structures make such systems very attractive [4, 5, 8].

As the MEMS/NEMS devices have been developed and applied, more and more frequently progress in their metrology has been also observed. Metrology is defined as the science of measurement including all practical and theoretical aspects. It is essential for scientific research and development (R&D) as well as for technological innovation. High quality metrology is critical to the major advances in all scientific fields. Moreover, metrology supports modern industrial competitiveness and development of new and improved products and processes.

Table 12.1 MEMS and NEMS devices-dimensions and mass

MEMS electromagnetically actuated cantilever		MEMS Silicon nitride microbeam	NEMS Silicon nanowires	Graphene based NEMS membranes (GNEMS)
			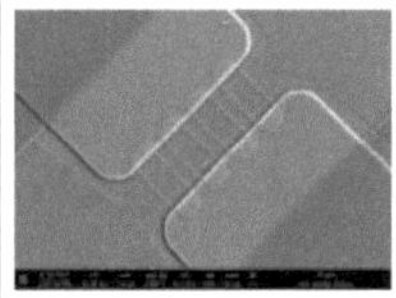	
Length (μm)	600	600	5	10
Thickness	10 μm	100 nm	50 nm	Non-defined
Mass	50 ng	50 pg	50 fg	Non-defined

Lord Kelvin quoted in 19th century: “If you cannot measure it, you cannot improve it”, which means that without reliable measurements, we do not understand properly and cannot control, manufacturing process in the reliable manner. Thus, advances in metrology and theirs effective use have a profound impact on our understanding of and ability to shape the world around us.

Moreover, it must be said that: not only is progress in nanotechnology, nanoscience and microsystem technology enabled due to the headway made in nanometrology but it also stimulates the development of measurement methods and techniques. This bidirectional relation makes the metrology research unique and of very high interest for almost all scientific fields.

Limitation of the today metrology is that the procedures, applied by National Metrological Institutes (NMIs) and introduced in the everyday practice, are mostly based on classical theorems (including classical elasticity theory, classical electrodynamics, thermodynamics etc.). It is clear, that because of the progress in miniaturization the role of the so called Quantum metrology will be increasing continuously. The first symptoms of this tendency, involving the introduction of the Josephson voltage, Hall quantum resistance reference standards, have been already indicated but it is evident to us that the more basic orientated research on this field will be necessary so that to make the mentioned theories accessible for wider group of scientist and technicians.

12.2 NEMS Sensitivity and Resolution

The vibrating MEMS/NEMS can be described basing on the simple harmonic oscillator (SHO) equation:

$$m\frac{\partial^2 x(t)}{\partial t^2} + \gamma \frac{\partial x(t)}{\partial t} + kx(t) = 0 \tag{12.1}$$

where m is the MEMS/NEMS mass, γ is the damping ratio and k is the MEMS/NEMS stiffness, x is the structure deflection. The above equation can be rewritten as:

$$\frac{\partial^2 x(t)}{\partial t^2} + \frac{\omega_0}{Q}\frac{\partial x(t)}{\partial t} + \omega_0^2 x(t) = 0. \tag{12.2}$$

where $Q = \frac{\sqrt{km}}{\gamma}$ is the quality factor and ω_0 is the angular frequency of the free structure vibration: $\omega_0 = \sqrt{\frac{k}{m}}$.

In many investigations MEMS/NEMS are used as resonators, which means that the deflection is excited by the external force, in this case:

$$\frac{\partial^2 x(t)}{\partial t^2} + \frac{\omega_0}{Q}\frac{\partial x(t)}{\partial t} + \omega_0^2 x(t) = \frac{f(t)}{m}. \tag{12.3}$$

where $f(t)$ is the force applied to a MEMS/NEMS to excite its vibration. The displacement $x(t)$ can be derived in frequency domain by Fourier transforming of Eq. (12.3):

$$X(j\omega) = \frac{F(j\omega)}{m}\frac{1}{\left(\omega_0^2 - \omega^2\right) + j\frac{\omega\omega_0}{Q}}, \tag{12.4}$$

where $F(j\omega)$ and $X(j\omega)$ are Fourier transforms of the force and MEMS/NEMS displacement respectively.

As investigations are usually performed using a network/spectrum analyzer in frequency space, the presented analysis in which MEMS/NEMS response is analyzed in frequency domain is very useful. Moreover, the recorded response signal can be split into real and imaginary or amplitude and phase components, which opens a field for a variety of detection techniques.

When Lorentz fitting is applied, the radial resonance frequency (ω_o), can be accurately determined. The full width at half maximum ($\Delta\omega$) is related to the damping coefficient γ. In all the mentioned above experiments the force F is related to the actuation technology, responsible for excitation of MEMS/NEMS movement. It's worth noticing, that the resonance frequency can be calculated as:

$$\omega_r = \omega_0\sqrt{1 - \frac{1}{2Q^2}}. \tag{12.5}$$

However, it should be also noticed that for high quality factors the angular frequency of free vibration is almost equal to the structure resonance frequency. Quality factor, which is the measure of the energy dissipation, influences the dynamics of the signal phase as well:

$$\varphi = \tan^{-1}\left(\frac{\omega\omega_0}{Q(\omega_0^2 - \omega^2)}\right). \tag{12.6}$$

12.2.1 Resolution and Sensitivity of MEMS/NEMS Devices

Two parameters of a MEMS/NEMS device are important to describe its properties: the sensitivity and the resolution. The sensitivity of a measuring system is the quotient of the change in an indication of a measuring system and the corresponding change in a value of a quantity being measured [29].

If a MEMS/NEMS device is applied as a mass change sensor the most efficient way to observe its sensitivity is to observe device resonance frequency, which is described by equation: $\omega_0 = \sqrt{\frac{k}{m}}$. Moreover, from technical point of view detection of small frequency changes can be performed with the highest precision, stability and accuracy using a modern phase locked loop (PLL) circuitry. If we differentiate the above formula the following formula describing mass change sensitivity can be obtained:

$$\frac{\partial\omega_0}{\partial m} = -\frac{1}{2}\frac{\omega_0}{m}, \tag{12.7}$$

which clearly shows that a higher mass change sensitivity is obtained when a MEMS/NEMS device of small mass exhibits high resonance frequency.

For the doubly clamped bridge, taking into account that its resonance radial frequency ω_{0b} and the effective mass m_b can be calculated basing on the following formulas: $\omega_{0b} = 2\pi \cdot 1.05\frac{t}{l^2}\sqrt{\frac{E}{\rho}}$ and $m_b = 2 \cdot 0.73wtl\rho$, the mass change sensitivity can be written as:

$$\frac{\partial\omega_{0b}}{\partial m_b} = 2.25\frac{1}{wl^3}\sqrt{\frac{E}{\rho^3}}. \tag{12.8}$$

with E, Young's modulus, ρ, mass density, w, bridge width and l, bridge length.

According to this formula the mass change detection sensitivity increases when factors $\sqrt{\frac{E}{\rho^3}}$ and $\frac{1}{wl^3}$ are maximized. In consequence materials forming a MEMS/NEMS device of high elasticity and low mass density are of big interest. In this case that the silicon carbide structures should exhibit 1.5 better sensitivity than the silicon devices. Moreover, devices of small length and width are preferable.

Similar analysis of the mass change detection sensitivity can be done for the supported beam-cantilevers, which leads to analogous conclusions. In general structures of low stiffness and high resonance frequencies are the most sensitive.

The performed above analysis indicates that the 2D NEMS structures are of big importance for the modern sensor technology. A NEMS device formed out of 2D material exhibits extremely high Young modulus, which results from the atomic structure of the vibrating components. In this case, if the structure does not contain any defects the 2D NEMS elasticity is the results of atomic bonds between atoms forming the movable NEMS part [2, 21].

The density of such a mechanical part is very low, as in the ideal case the movable components are built only out of atoms. Although, at this place difficulties in the interpretation of the observed phenomena, especially taking into account classical theories of the mechanical and electrical engineering, must be noticed.

In general, if an elastic structure is being deformed by an external force, it experiences internal resistance to the being induced deformation and restores it to its original state if the external force is no longer applied. There are various material parameters: the so-called elastic moduli, such as Young's modulus, the shear modulus, and the bulk modulus, which describe the inherent elastic behavior of a structure as a resistance to deformation under an applied load. The mentioned moduli apply to different kinds of deformation, e.g. Young's modulus applies to extension/compression of the body, whereas the shear modulus applies to its shear.

The elasticity of a structure is exactly described by a stress-strain characteristic, which is the relation between the so-called stress (which is the ratio of average restorative internal force over unit area) and strain (which is the relative structure deformation). In general, the characteristics is nonlinear, but by use of a Taylor series it can be approximated as linear for sufficiently small deformations (which means, that the higher-order terms can be neglected).

The linearized stress-strain relationship is described by the Hooke's law, which is usually applied to the elastic limit for most metals or crystalline materials. It must be noted additionally, that because in general stresses and strains can have various independent components the proportionality factor can no longer be just a single number but rather a tensor containing real numbers (e.g. defined for various crystallographic directions). Moreover, Young's modulus is the technical parameter resulting from interaction in continuum structure: it is therefore an average of various atomic and molecular interaction which internally occur in the being deformed body.

This classical approach can and is very often applied in the MEMS analysis, which are big enough to be described with global parameters like e.g. Young's modulus. However if the analysis of NEMS is considered, the fundamental questions arise if this approach can be applied for structures, whose thickness corresponds to the atoms or molecules which can be sometimes even counted. The most exact approach is in this case to calculate elastic properties in ab initio, in other words *from the first principles*, manner. An ab initio calculation starts in this case with the properties of constituent atoms forming e.g. graphene flake and the laws of quantum mechanics. In general the characteristics of isolated individual molecules can be derived and consequently followed by computations of the interactions of larger and larger groups of molecules and as a result the properties of the entire NEMS structure can be calculated. When real structures are taken into consideration, e.g. polycrystals, containing defects, supported or suspended on bulk holders, the ab initio calculations are extremely difficult, time and labor consuming. The obtained results are quite often not the intuitive ones and from the NEMS design and interpretation point of view not helpful.

Moreover, in MEMS and NEMS analysis it is essential that the terms stress and strain must be defined without ambiguity. If NEMS are considered, the structure deflection is usually of fractions of nanometers and is observed quite often in the frequency bandwidth of up to 100 MHz. This creates a lot of technical problems in the deflection detection and if the metrology is considered the procedures are even more difficult.

The second parameter describing metrology using a MEMS/NEMS device is its resolution described as the smallest change in a quantity being measured which causes a perceptible change in the corresponding indication [29].

The analysis of the MEMS force measurement resolution illustrates the general rules governing the way how to improve the system parameters. In the MEMS/NEMS technology the static force is observed by the detection of the spring beam static deflection. In general stochastic vibration of the MEMS/NEMS molecules and the vibration of the molecules around a system form background of the force interaction observations. In analogy with electrical engineering, when one assumes that the force is the analogy to the electrical voltage and MEMS/NEMS speed corresponds with the electrical current the minimum detectable force $F_{\min}$ can be calculated like Johnson-Nyquist noise:

$$F_{\min} = \sqrt{4k_B T \gamma B}, \tag{12.9}$$

where k_B is the Boltzman constant, T is the MEMS/NEMS temperature, B is the measurement bandwidth. If we notice that $\gamma = \frac{k}{\omega_0 Q}$ and assume that Q is high enough, which makes that ω_r and ω_0 are equal, the formula for the minimum detectable force is:

$$F_{\min} = \sqrt{\frac{4k_B kTB}{Q\omega_0}} = \sqrt{\frac{2k_B kTB}{\pi Q f_0}}, \tag{12.10}$$

where f_0 is the free vibration frequency. It can be clearly seen, that the cantilevers of small stiffness and high free vibration frequency exhibit higher resolution of force detection. For the MEMS cantilever presented in Table 12.1 with the stiffness of 15 N/m, resonance frequency of 50 kHz, the smallest detectable force is 42 $\text{fN/Hz}^{0.5}$ in the bandwidth of 1 Hz but for the silicon NEMS cantilever with 20 μm length, 0.3 μm thickness and 10 μm width, corresponding with stiffness of 1.7 N/m and resonance frequency of 320 kHz, the minimum detectable force can be reduced to 7 $\text{fN/Hz}^{0.5}$ (both parameters were calculated for quality factors of 300 and ambient temperatures). When the measurements are done in liquid nitrogen (LN) and vacuum, which usually leads to quality factor increase to 10,000, the minimum detectable force is reduced to 60 $\text{aN/Hz}^{0.5}$.

Another very important issue, illustrating the capabilities of the NEMS devices is the small operating power $P_{\min}$ needed for their activation. For a resonant MEMS/NEMS power $P_{\min}$ can be calculated by noticing that the device is a lossy energy storage. Energy injected into the device is dissipated in a time:

$$\tau \approx \frac{Q}{\omega_0} \tag{12.11}$$

which is often named as the ring-down time of the resonator. The minimum operation energy for the system is the energy, which will drive the NEMS at amplitude corresponding with the thermal fluctuation. Given the energy $k_B T$ of the mode thermal fluctuation the minimum power $P_{\min}$ can be estimated in this case as:

$$P_{\min} \approx \frac{k_B T}{\tau} = \frac{k_B T \omega_0}{Q}. \tag{12.12}$$

For a NEMS device which can be fabricated using today's technologies the minimum power $P_{\min}$ is on the order of tens of aW (10^{-17} W). One can estimate that even for the systems operating with arrays containing thousands of NEMS resonators, the minimum power needed for the operation is far smaller than for the electronic devices, in which power of several microwatts is usually dissipated.

12.2.2 NEMS Fabrication Technology

The vast majority of technologies applied in fabrication of NEMS devices is based on microfabrication procedures applied in the manufacturing of microelectronic integrated circuits (ICs). A lot of various techniques including etching, lithography, thin film deposition, resist processing are available in the research laboratories. However complexity and difficulties, connected mostly with the fabrication of structures whose dimensions are in the range of tens of nanometers, must be mentioned. To the most time consuming, cost and equipment involving processes belong electron beam (E-beam) lithography, deep silicon etching and integration with application specific integrated circuits (ASICS). The described above methodology can be applied in single wafer or batch processing and the latter one is the cost-effective method of dealing with volume manufacturing. One of the ways enabling manufacturing of the prototype NEMS is to fabricate the device out of thin film deposited on solid substrate e.g. silicon wafer. In this way application of costly silicon on insulator (SOI) or silicon carbide (SiC) substrates as well as expensive deep silicon etching processes will be avoided [18].

In the proposed technology silicon nitride NEMS bridges were fabricated [25]. In order to enable detection and actuation of NEMS vibration metallization paths were integrated with moveable components. In general fabrication of silicon nitride films of thickness less than 200 nm is difficult due to high tensile stress occurring while thin film deposition. As the result the adhesion between silicon nitride and silicon is small, leading to the structure instability. However, by the process parameters optimization (changing of the substrate temperature, gas flow, gas ratio, pressure and time) it was possible to produce thick silicon nitride and deposit a layer resist for wet etching in KOH solution.

The entire fabrication process of the NEMS bridge consists of three steps. In the first fabrication step a stoichiometric silicon nitride Si_3N_4 layer (200 nm in thickness) is deposited in LPCVD method with (SiH_2Cl_2/NH_3) gas ratio of 1:4. The stoichiometric LPCVD nitride shows a very large tensile stress (~1.7 GPa) and is compensated by a subsequent non-stoichiometric silicon nitride deposition. The non-stoichiometric silicon nitride SixNy (200 nm in thickness) was deposited with SiH_2Cl_2/NH_3 gas ratio of 8:1. The conducted process results in the reduction of the stress in the deposited layer forming a NEMS bridge to ca. 0.2 GPa. In the second fabrication step chromium (as an adhesion layer) and platinum layers were sputtered. In a dry plasma etching process of the platinum and the silicon nitride the metallization paths and shapes of the NEMS bridge are defined. Finally the structures are released by wet etch silicon in KOH solution. A variety of NEMS bridges with thickness of up to 500 nm, width of 50 μm and length of 420 μm were fabricated in this scenario. All the structures were wire-bonded to the printed circuit boards (PCB) which made it possible to detect and actuate structure vibrations. The silicon nitride bridge, fabricated in the above described technology, of dimensions: length 280 μm, width 40 μm and for the Young modulus of 160 GPa, and silicon nitride density of 2700 kg/m^3 exhibits mass change density calculated according to Eq. (12.8) of 1.24 MHz/ag.

It must be noted however, that even if the NEMS fabrication process is simplified it is still challenging. According to the state of the art, NEMS solutions are based on electron beam (E-beam) lithography and following quite complicated etching steps associated with depositions and photolithography procedures. Such an approach is extremely expensive in terms of process complexity, characterization, instrumentation and very time-consuming. Moreover, in this way fast prototyping needed, when the possibilities of the NEMS tool should be assessed are very limited, as in the standard process only the wafer scale manufacturing is possible. One of the solutions to these problems is the application of focused ion beam (FIB) and scanning electron microscopy (SEM) based techniques. In the standard approach, the FIB and SEM technologies, forming the so called dual platform systems (integrating both ion and electron beam columns), have been widely used for microfabrication, nanofabrication and testing of the IC's. The focused gallium ions and electrons are directed onto the samples and applied to modify or image the investigated sample. In this way FIB and SEM tools can sputter and implant lines as narrow as 20 nm and deposit metals and insulators in lines as narrow as 20 nm in user-defined patterns.

The focused electron beam and scanning electron microscopes have been extensively used for the fabrication of the transmission electron microscopy (TEM) sample preparation. In this case the beam of gallium ions carry enough energy to make the sample so thin that it will be transparent for the electrons in a transmission electron microscope. What must be underlined is that the process can be controlled with precision and resolution of fractions of nanometers. That is the reason why both technologies have attracted much interest in the fabrication of NEMS devices. The most straightforward procedure is to use highly energetic gallium ions for modification (milling down) MEMS devices. In this way sensors for scanning near field optical microscopy (SNOM) were successfully fabricated by opening an aperture of tens of

Fig. 12.1 NEMS silicon nanobridge fabricated using FIB technology; silicon bridges were patterned with gallium ion beam and wet etched; bridge thickness ca. 30 nm; estimated device mass 10 fg

nanometers opening in the microtip [10]. The FIB milled SNOM nanoprobes are only an example of the application of the described technology in the nanotools experiments. There are other examples, in which MEMS as basic structures were applied and by the milling done at the defined positions nanotip cantilevers for advanced scanning probe microscopy (SPM) or tiny calibration references were fabricated.

Besides ion milling there is another nanofabrication technology which relies on galium ion FIB implantation. In this case when on a silicon crystal sample a beam of galium ions is directed the implanted areas form a structure which is resistant to subsequent etching in KOH solution (in other words the implanted areas are an etch stop structures in wet etching procedure). The etch stop technology has been known in microelectronics for almost 2 decades but it was applied in the FIB technology by Murano group for the first time [20]. In this way 3 dimensional (3D) structures can be created, whose dimensions are of tens of nanometers. In Fig. 12.1 a set of galium implanted silicon nanobridges is presented. As it can be seen, the silicon crystal walls are clearly visible which is typical for the wet etching process. The nanobridge structures are of 30 nm thickness, which makes the structure almost transparent for the electrons used for SEM imaging. The estimated nanobridge mass is of 10 fg, which is almost three orders of magnitude smaller mass then the mass of silicon nitride nanobridges presented in Table 12.1.

The described technology makes it possible to fabricate a variety of NEMS and is very attractive for the single devices or prototype investigations. It was also utilized in the fabrication of the array NEMS [19].

The FIB/SEM technology enables also fabrication of 3D nanostructures in the so called focused electron beam induced deposition (FEBID) process. In this scheme the investigated structure is immersed in environment of metal organic (MeO) precursor. On the areas exposed to the electron beam, metal (Me) containing structures can be deposited. In contrast to FIB milling, which is the top-down procedure (a manufactured structure is fabricated out of the larger part) FEBID process is the bottom-up procedure (which means that the particular components are being integrated to give rise to more complex systems).

The FEBID technology is a mask-less deposition technique suitable for NEMS rapid prototyping. The deposition process can be performed faster and in the more flexible way than in conventional microelectronics fabrication processes. As the dimensions of the fabricate devices correspond to the resolution of a scanning electron microscope the metal containing structures can be placed with resolution of tens of nanometers. There is a variety of metal containing precursors which are used in FEBID technology. The conductive wires are fabricated basing on platinum, gold, cobalt or silver [3, 11, 16, 30]. FEBID technology makes it not only possible to fabricate high frequency (HF) resonators but also devices or components for NEMS deflection detection. In this case solutions based on field emission (FE) and piezoresistivity can be basically distinguished (further details can be found in the chapter on NEMS deflection detection methodology).

An important concept of the metrology is the traceability, which means that the result of a measurement must be related to stated references through an unbroken chain of comparisons with stated uncertainties. In case of the above-described NEMS-deflections investigations, there is lack of reliable and easily accessible calibration routines which leads to discrepancies in the assessment of various experiments. In this way FEBID technology was applied to verify the NEMS mass change detection resolution, which can be estimated using Eq. (12.8). To perform such an experiment NEMS device must be loaded with the mass which is known and the resonance frequencies before and after mass loading must be measured. In Fig. 12.2 platinum/carbon FEBID lines deposited on a silicon nitride bridge (Table 12.1) are visible. As it can be estimated from atomic force microscopy (AFM) image the line height and line width is of 50 and 20 nm respectively (see Fig. 12.3). In this way the mass of ca. 7 pg can be calculated which leads to the resonance frequency shift of 1324 Hz. The experimentally verified mass change detection resolution is of 0.18 Hz/fg. The differences between theoretically calculated and verified parameters stem from many reasons and illustrate very well the problems of nanometrology performed with NEMS tools. It should be noted that, Young's modulus of the silicon nitride and thin film metallization films cannot be described precisely and are technology and process dependent. Moreover, shape of the bridge anchoring points, due to the crystal structure of the silicon substrates, make elasticity calculations very difficult or almost completely impossible. The density of platinum/carbon FEBID reference structures depends on the way how the FEBID process is performed. Despite all these difficulties, the proposed routine makes it possible to confirm the sensitivity of the mass change detectors based on silicon nitride bridges.

As it was mentioned previously besides the sensor sensitivity the second parameter describing a NEMS device is its resolution. In the simplest approach, the resolution of the resonance frequency corresponds with the resolution of the system using which the vibration amplitude or phase are observed. The higher resolution of the mass change detection can be achieved when a NEMS resonator vibrates with high quality factor Q. In this case the bandwidth of the resonance frequency Δf can be estimated using equation $\Delta f = f_0/Q$, and for typical silicon nitride bridges

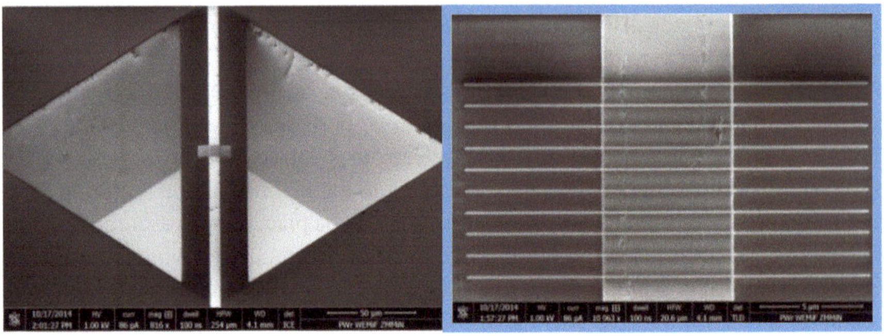

Fig. 12.2 NEMS silicon nitride bridge with FEBID deposited structure for mass-change detection-sensitivity calibration

Fig. 12.3 AFM image of the FEBID calibration structures used to calibrate the mass change sensitivity

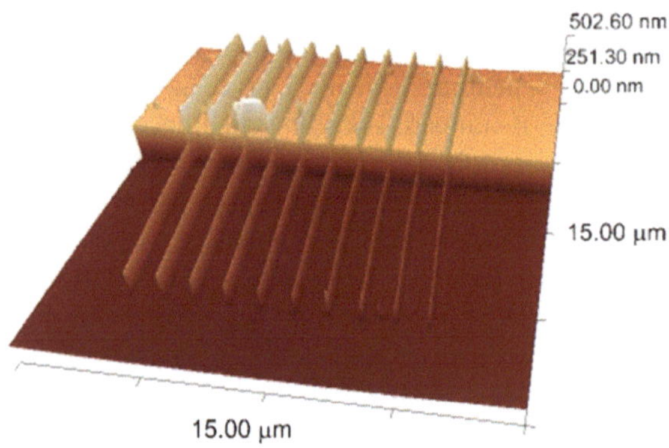

presented in Table 12.1 it is of hundreds of hertz for measurements performed in ambient conditions. The higher quality factors can be achieved when the dissipation in the entire NEMS structure is reduced, which involves careful engineering of the bridge, its anchoring areas and metallization lines. Application of the so called quality factor enhanced technology, which in fact is the setup with the positive feedback loop, makes it possibly to decrease the bandwidth of the resonance frequency even further [36]. In this case however, an integration of a deflection actuator with the mechanical beam is required so that to control the structure deflection. In order to observe the resonance frequency various frequency detector can be used. The most efficient way to observe the resonance frequency change is the application of the so called phase locked loops (PLLs) systems. In the PLL systems the signal corresponding with the bridge vibration is compared with the reference signal of a digital quartz generator. The output frequency is controlled in a feedback loop so that both frequencies are equal. The resolution of the PLL detector, limited by the stability of the reference signal generator, is usually around tens of millihertzs.

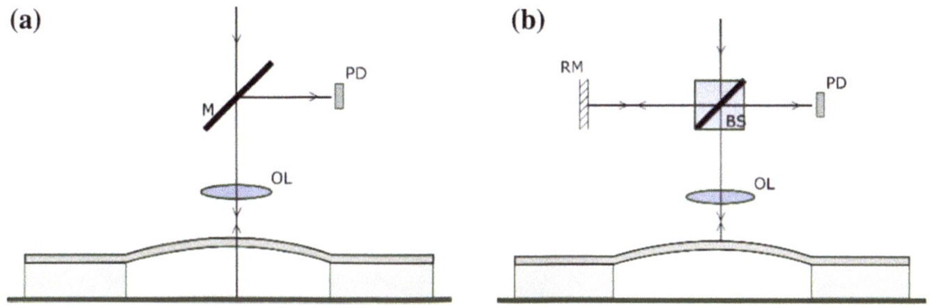

Fig. 12.4 Optical detection of NEMS deflection: **a** Michelson interferometer, **b** Fabry-Perot interferometers; *M* interferometer mirror, *OL* objective lens, *BS* beam splitter, *PD* photodetector

12.3 NEMS Detection and Actuation

12.3.1 Optical Methods

There are many approaches, making it possible to observe the NEMS displacement using optical technology. In the laboratory experiments setups basing on interferometric bulk design are often applied. In particular path stabilized Michelson interferometry and Fabry–Pérot interferometry, have been used in such measurements. In Michelson interferometers, a focused laser beam reflects from the surface of a NEMS device and interferes with a stable reference beam. In turn, in Fabry-Perot interferometers the optical cavity formed between the vibrating NEMS and its substrate modulates the optical signal on a photodetector when the nanoresonator deflects in the out-of-plane direction. In this way shot noise limited displacement sensitivities of ca. 10^{-6} nm/Hz$^{0.5}$ Hz are routinely attainable on objects with cross sections larger than the diffraction limited optical spot in the bandwidth up to 50 MHz.

In Figs. 12.4 and 12.5 a measurement setup comprising a Michelson interferometer and the holder with a mounted silicon nitride NEMS bridge is shown. In Fig. 12.6 the resonance curve of NEMS structure is presented. A piezoelectrical actuator was used for excitation of resonance vibration of 1.5 nm amplitude. Basing on the performed experiments resonance f_0 frequency of ca. 690 kHz can be identified. The quality factor Q of the vibrations in air can be estimated of ca. 170 (Fig. 12.6).

The results obtained in this way are the quantitative ones, which is, besides the high resolution, the second advantage of the interferometrical technology. Thus, not only device diagnostics but also characterization and study of the NEMS properties can be done with high precision. The most important drawback of the described solution is related to the fact that the bulk optics, due to big mass and dimensions, cannot be integrated with the fabricated NEMS devices. From that point of view optical fibre setups are of much interest for NEMS technology. The small diameter of the optical fibre makes it possible to integrate the probing sensor i.e. optical fibre directly with a NEMS device. Such displacement sensors are often built in fibre Fabry-Perot interferometer (F-FPI) configuration in Fig. 12.7. In this setup an optical fibre is the

Fig. 12.5 Measurement setup using a bulk interferometer for silicon nitride NEMS bridge metrology

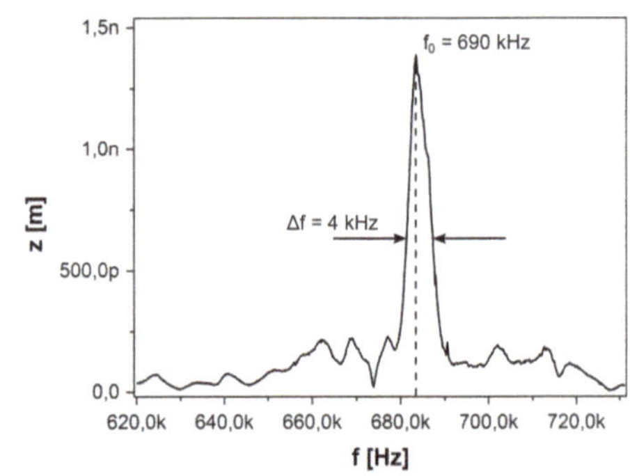

Fig. 12.6 Resonance curve of silicon nitride NEMS bridge

central part of the sensing system. A laser light source is connected to the arm *A* of the fibre coupler. The NEMS device is illuminated with the beam coming out from the arm *B* in Fig. 12.7. Part of this beam is reflected from the NEMS object and is coupled back to the fibre. Due to the difference between the refraction indexes of the air and optical fibre part of the beam emitted by the laser source is reflected from the arm *B* open end. The light beams, which are reflected from the measurement object and from the fibre open end interferes together and the interference fringes

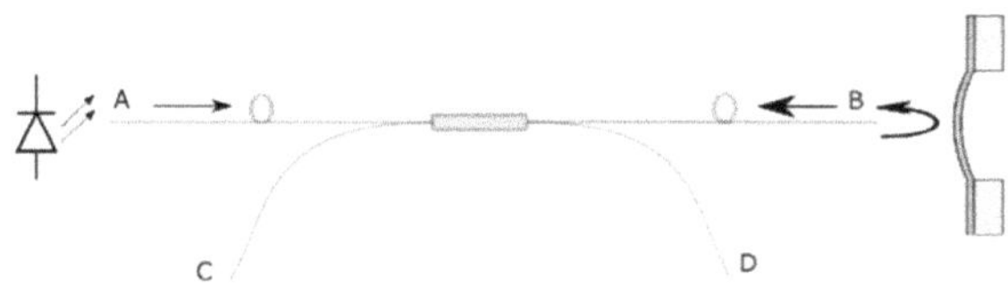

Fig. 12.7 Fibre Fabry-Perot interferometer for NEMS deflection investigations

are observed with a photodetector connected to the arm *C* of the fibre coupler. As it can be seen the system is very compact and enables observation of the deflection of many NEMS devices in parallel. The proposed technology makes it possible to observe NEMS deflection with the resolution of 30 fm/Hz$^{0.5}$ [34].

If instead of a laser diode a noncoherent light source is used, the photodetector connected to the arm *C* detects the intensity of the beam reflected from the moving object in Fig. 12.7. In this case the photodetector signal corresponds with the distance between the optical fibre and the NEMS device. The resolution of the so called light intensity fibre optic sensor (I-FODS) is 120 pm/Hz$^{0.5}$. In comparison with the F-FPI the I-FODSs exhibits smaller resolution but can operate in the wide frequency bandwidth of up to 1 MHz. Moreover the output signal of a I-FODS is continuous which simplifies signal acquisition and data analysis, as the F-FPI requires quite cumbersome signal processing [27].

Besides optical technologies electronic systems for the detection and measurement of NEMS deflection play a very important role. Integration of a NEMS device with an electronic measurement and control environment makes it possible to develop solutions for not only laboratory but more practical applications as well [28]. Moreover, integration of the measurement and control functions in application specific integrated circuits (ASICs) enables fabrication and usage of the NEMS based devices in portable systems whose characteristics is precisely and repeatable described. The full strength of the NEMS technology can only be utilized, when it is possible not only to sense but to control the structure vibration. To control the NEMS vibration means in this case to actuate static and/or resonance structure deflection and to maintain it at the defined level with a defined speed and frequency range.

As the size of a NEMS device is being reduced and as a consequence of this operation, frequency increases the modulation of the impedance of the deflection sensing and actuation components is getting smaller. Simultaneously, the influence of stray or parasitic impedances increases which makes detection of the tiny effects very difficult. NEMS detection and actuation electronics quite often operates on its limits, which means that the electrical signals connected with the NEMS deflection are comparable with the input thermal noise of an applied amplifier and/or its bias current. Below a summary of various methods applied for the sensing and actuation of NEMS devices is presented.

12.3.2 Piezoelectric Technology

In piezoelectric materials, a mechanical stress induces electrical charge on the device electrodes. Alternatively, an applied voltage (in other words charge deposition) induces a mechanical stress in the biased crystal, which leads, depending on the elastic structure properties, to the strain of the piezoelectric system. The described technology can be used to detect and drive mechanical motion in the frequency bandwidth of up to 1 GHz. The transduction efficiency is very high and this technology is often used in MEMS devices. However, in NEMS technology the main difficulty relies in the fact that the crystal integrity must be maintained even for the structure of nanometer thickness. From that point of view, having in mind that the piezoelectric actuator must be resistant to the electrical breakdown, integration of the piezoelectric technology with NEMS devices of thickness smaller than 100 nm is quite difficult [1]. In the laboratory experiments piezoelectric actuators are very often applied to test the structure resonance properties, however in this case control of the structure vibration is difficult and cannot be done in the repeatable manner.

12.3.3 Capacitive Technology

The capacitive technology has been carried over from MEMS technology. An attractive force occurs between the plates of a capacitor when its electrodes (plates) are charged. In order to induce electrostatic force, one usually fabricates a gate electrode around the NEMS device. The most convenient and the most efficient way to excite the electrostatic force is to bias the substrate under the NEMS resonator. Many MEMS technologies can be introduced on this field, however it must be noted that stray capacitances around the NEMS device are much bigger than the capacitance between the substrate and the NEMS resonator. As the consequence the capacitive technology is less effective for higher frequencies, however solutions making it possible to control the structure deflection at frequencies of 700 MHz were reported [33]. In capacitive displacement technology, the motion of the mechanical element of a NEMS device modulates the electrical capacitance between the moving NEMS part and a fixed electrode. To detect this capacitance modulation, the NEMS capacitor is usually biased with a voltage and the current is detected using a current to voltage (I/U) converter. The biggest problem in this case is that the stray capacitances around NEMS resonator and in the I/U converter are even one order of magnitude bigger than the capacitance which is being detected. One solution to the problem is the application of a transformer ratio bridge, whose inductive arms negate the capacitive background [12, 35]. Another technology relies on implementation of the switched amplifiers using which it is possible to take into account the stray capacitances and detect very feeble signal corresponding with NEMS capacitance.

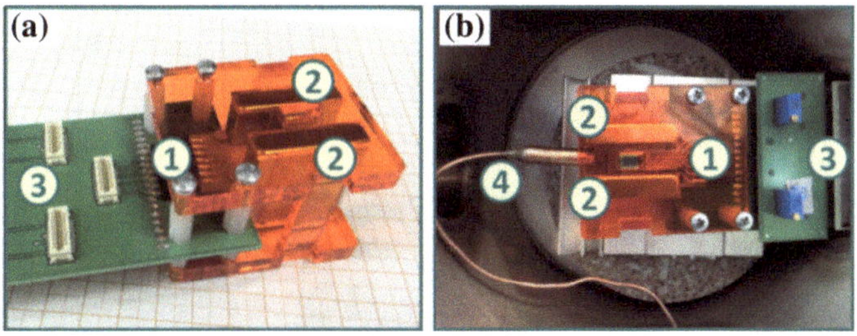

Fig. 12.8 Experimental setup of a silicon nitride bridge-electromagnetic actuation and detection. **a** Side view, **b** top view. *1* Silicon nitride bridge mounted on a printed circuit board (PCB), *2* neodymium magnets, *3* measurement and control electronic, *4* sensor for magnetic field measurements

12.3.4 Electromagnetic Technology

In the presence of a magnetic field a NEMS device deflection can be both detected and actuated by coupling of the magnetic field with a current carrying loop integrated with the NEMS device. This method is known as the electromagnetic or magnetomotive actuation and detection and can be used to drive vibration at even microwave frequencies. The drive force of a current I carrying conductor in magnetic field B is called Lorentz force F_L and can be calculated using the following formula:

$$\overrightarrow{F_L} = L_L \overrightarrow{I} \times \overrightarrow{B}. \tag{12.13}$$

where L_L is the length of the conductor. Early experiments were carried out in strong magnetic fields of few teslas generated in superconducting magnets, which made the measurement setup expensive and very large [6, 13]. Progress in fabrication of permanent magnets has made it possible to design setups of relatively small dimensions in which magnetic field of fractions of tesla are generated. In Fig. 12.8 a silicon nitride bridge with a metallization line of resistance of 120 Ω placed between two neodymium magnets generating field of 0.5 T is presented. In the presented configuration Lorentz force of 140 nN can be estimated for the silicon nitride bridge of 280 μm length, drive current of 1 mA and magnetic field of 0.5 T. The estimated Lorentz force is big enough to induce NEMS structure vibration of tens of nanometers vibration amplitude. In Fig. 12.9 resonance curve of a silicon nitride film measured in vacuum of 10^{-4} Pa is presented. As it can be seen the vibration amplitude of tens of nanometers can be relatively easily measured using a Michelson interferometer. In this way static bridge deflection, which means deflection of frequency much smaller than the resonance bridge frequency, in the range of fractions of nanometers can be induced and precisely controlled.

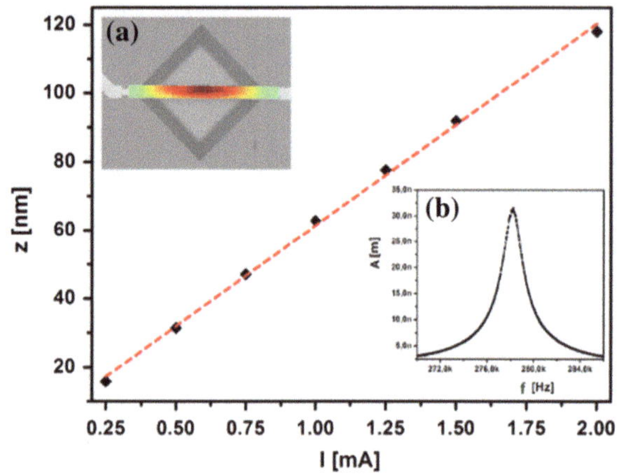

Fig. 12.9 Electromagnetically actuated resonance silicon nitride bridge vibrations

Electromagnetic technology was applied to detect the NEMS vibration for the first time in 1999 [7]. The technique is based upon the presence of a uniform static magnetic field, in which a NEMS containing a conductive wire vibrates. The time varying magnetic flux generates an induced electromotive force (EMF). For a doubly clamped bridge an EMF is given by:

$$EMF = \xi B L_L \frac{\partial x(t)}{\partial t}. \tag{12.14}$$

where ξ is a geometric factor (0.885 for a doubly clamped beam). For the beam, whose measurements were presented in Fig. 12.10 one can calculate, that the vibration period is ca. 3.5 μs. If one assumes that the vibration amplitude is 40 nm at frequency of 280 kHz, the average speed of the bridge is 44 nm/μs. For the structure length 280 μm and magnetic field 0.5 T EMF is ca. 5.5 μV. The measurement of such a low voltage is quite difficult, but having in mind that the thermal voltage noise of modern instrumental amplifiers is of few nV/Hz$^{0.5}$ a careful engineering makes it possible to record a signal corresponding with the resonance NEMS vibration.

The electromagnetic actuation and detection of NEMS vibration can be combined, which makes the described technology extremely interesting. When a Lorentz loop is biased the structure can be excited to vibration. When it starts to oscillate in resonance a EMF voltage can be identified across the conductor. The results of the simultaneous vibration and bridge detection are presented in Fig. 12.10a, where a family of resonances was recorded for various bridge bias currents. In Fig. 12.10b the resonance curve and the relevant phase change curve are illustrated [22].

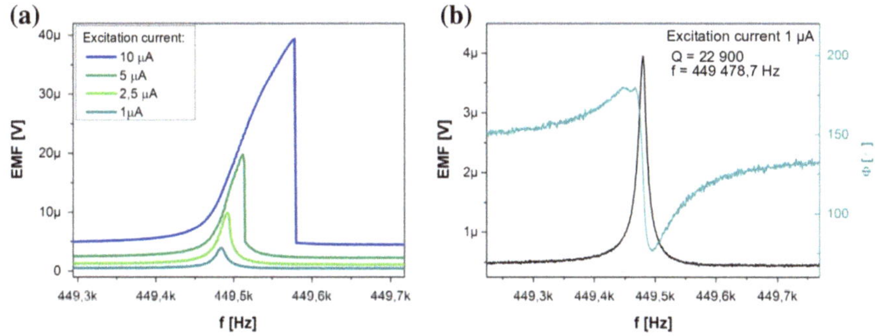

Fig. 12.10 Electromagnetic actuation and detection of electromagnetically driven and detected NEMS vibration, **a** resonance curves recorded for various drive currents, **b** resonance curve with phase shift changes

12.3.5 Piezoresistive Technology

In general piezoresistors are the sensing structures whose resistance is the function of strain. Changes of the resistance *R* can be described as:

$$\frac{\Delta R}{R} = \left[(1 + 2\vartheta) + \frac{\Delta\rho}{\rho}\right]\varepsilon = \mathrm{GF}\varepsilon. \tag{12.15}$$

where ΔR and R are the change in resistance and total resistance respectively, ε is the applied stress, ν is the Poisson's ratio, $\Delta\rho$ and ρ is the piezoresistor resistivity respectively. The coefficient *GF* is the so called gauge factor, which describes the sensitivity of the strain sensing structure. When the piezoresistor is made out of metal its response is the function of the geometry changes and the term $\frac{\Delta\rho}{\rho}$ can be neglected. For metal piezoresistors *GF* ranges from 1.5 to 3.5 depending on material alloy. In the semiconductor piezoresistor the resistivity change $\frac{\Delta\rho}{\rho}$ becomes dominant and the geometry related term $(1 + 2\vartheta)$ is usually not taken into account. In this case strain induces changes in the conduction band Fermi level and as the consequence the gap size is modulated.

The piezoresistive detection scheme can be also implemented in NEMS technology. The resonator itself may be manufactured from a piezoresistive material or alternatively a piezoresistive material can be integrated with the resonator. Typically this is done at the clamping points of the resonator, where the strain is the highest. As the piezoresistive technology is the electrical one it can be easily implemented into a heterodyne down-mixing circuit where the resonant response frequency is multiplied with a bias frequency to obtain a low frequency detection signal, whose acquisition is not sensitive to parasitic capacitances. This method has been successfully applied for mass sensors and demonstrated high resolution in scanning probe microscopy (SPM) investigations [9, 31, 37]. As NEMS dimensions shrink, integration of a deflection sensing device with the nanomechanical structure is becoming challenging. Besides, solutions based on nanoelectronics FEBID techniques can be of a real support. The FEBID nanostructures are often described as composites consisting of metal nanograins embedded in non-conductive matrix whereas electron transport has been defined as inelastic tunneling process occurring among nanograins [14, 24]. This phenomenon makes it possible to apply these structures as sensors of strain, which is induced by deflection of the device integrating FEBID nanogranular piezoresistors (NGRs). The induced strain results in the modulation of the distance between embedded in an amorphous matrix metal nanograins. In this way the probability of electron tunneling among the nanograins is modulated, which in turn is observed by the changes in the resistance of the FEBID structure. As the tunneling phenomena are the quantum mechanical ones, it is expected that the deflection detection sensitivities should be very high. However, several measures must be undertaken in order to ensure stability and response repeatability of the FEBID piezoreistors. If an organometallic platinum precursor is used the fabricated FEBID piezoresistor consists of conductive platinum nanocrystallites immersed in carbon containing matrix which results in high resistivity of the deposited material and making high frequency (HF) measurements difficult. Moreover, carbon can easily oxidize in ambient conditions, which strongly influences carrier transport stability. Another important problem is that of process repeatability as the beam parameters, chemical composition and stability of the chemical precursors used for the deposition may differ from experiment to experiment [17].

The FEBID NGR structures were applied as sensors to detect the micromechanical cantilever deflection [32]. Basing on the microcantilever vibration analysis it is possible to use such structures as a mass change sensing platform. When the mass change sensitivity is concerned it is well known that the smaller platform mass corresponds with the increased mass change sensitivity. Therefore, the proper way to detect smaller mass change is to fabricate structures of smaller mass like e.g. silicon nitride bridges. However, it has to be noticed, that the smaller structures vibrate at higher frequencies and their oscillation amplitude is much smaller than that of the MEMS microcantilevers. The described features define requirements for the components detecting mechanical structure vibration, which have to exhibit high sensitivity and have to operate in wide frequency bandwidth. To the possible solutions to the presented problems belong again the FEBID technology, which ensures the high sensitivity and makes it possible to fabricate small vibration detectors.

Mechanical structure of a silicon nitride bridge is shown in Fig. 12.11. Accordingly, the following formulas describe its elasticity:

$$R_A = R_B = \frac{F}{2}, \tag{12.16}$$

$$M_A = M_B = -\frac{1}{8}FL, \tag{12.17}$$

$$f_{\max} = \frac{FL^3}{192EI_z} = \frac{FL^3}{16Ewt^3}, \tag{12.18}$$

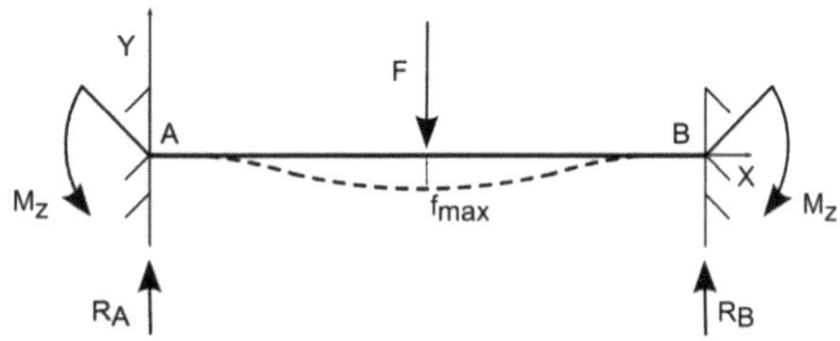

Fig. 12.11 Mechanical structure of a silicon nitride NEMS bridge

where L, w, t are the length, width and thickness of the bridge, F is the force acting in the middle of the structure, E is the Young modulus.

Stress at the bridge supporting points can be calculated using equation:

$$\sigma_{\max} = \frac{Mt}{2I_z} = \frac{3}{4}\frac{FL}{wt^2}, \tag{12.19}$$

where I_z is the inertia modules of the beam: $\mathrm{I_z} = \frac{\mathrm{wt}^3}{12}$. Strain of the microbridge at the beam supporting points can be calculated basing on the Hooke's law:

$$\varepsilon_{\max} = \frac{1}{E}\sigma_{\max} = \frac{3}{4}\frac{FL}{Ewt^2}. \tag{12.20}$$

By combining the above equations, it is possible to calculate the strain of the microbridge, when the maximal deflection is known:

$$\varepsilon_{\max} = \frac{12 f_{\max} t}{L^2}. \tag{12.21}$$

Therefore, relative change of the piezoresistor resistance deposited in FEBID technology at the beam supporting points is:

$$\frac{\Delta R}{R} = GF\varepsilon_{\max} = GF\frac{3}{4}\frac{FL}{Ewt^2}. \tag{12.22}$$

Application of the above equation is limited as Young's modulus of the entire bridge structure is unknown and difficult to be determined. This results from the fact that the properties of silicon nitride substrate depend on the deposition process. Moreover, the silicon nitride substrate is covered with a platinum thin film, which makes determination of the effective Young's modulus more cumbersome.

Usually, in order to obtain better sensitivity of the structure bending and to concentrate the stress in the area of the NGR piezoresistors the FIB milling will be conducted. The FIB milling is performed in close vicinity of the clamping of the microbridge in order to decrease the width of the bridge exactly in the area, where the NGR piezoresistor should be located. Shape of the FIB modified area is designed in the way to ensure the homogenous distribution of the stress along the NGR piezore-

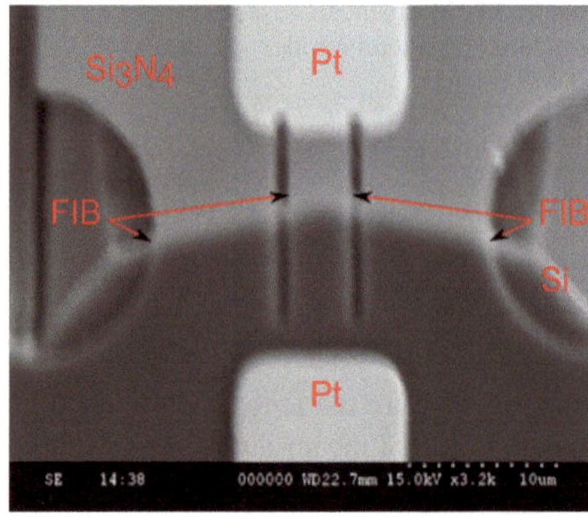

Fig. 12.12 FIB processing of the bridges prior to the FEBID piezoresistors deposition

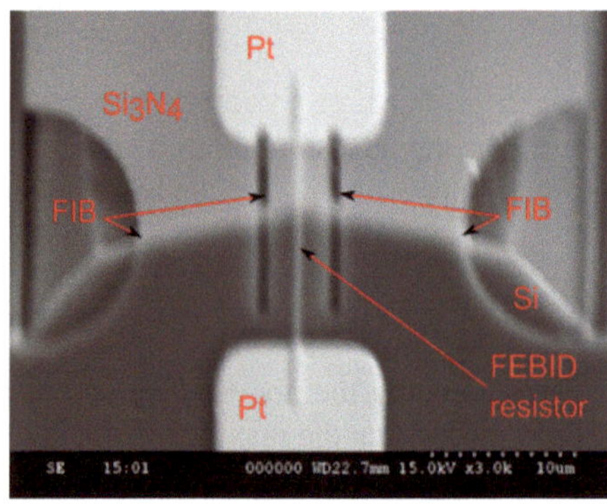

Fig. 12.13 NEMS bridge after the FIB modification with deposited FEBID resistors

sistor. As the deflection sensors receive the entire stress associated with the structure movement, their response should be the highest.

In Fig. 12.12 NEMS bridge structure with the FIB fabricated openings is shown. The FIB processed structures at the bridge supporting points and in the middle of the structure, where the FEBID piezoresistors are deposited. The FIB milling procedure was conducted in the selective way, so that to modify only the silicon nitride layer-in Fig. 12.13 the silicon substrate is clearly visible. In Fig. 12.14 the NEMS bridge after the FIB modification with the FEBID deposited piezoresistors is shown.

In order to measure the influence of the performed modifications the bridge resonance frequency were measured using a Michelson interferometer. The resonance frequency of the NEMS bridges prior to the modifications was 791 kHz. After the FIB processing the resonance frequency is decreased by around 80 kHz, which indicates decrease in the structure stiffness of ca. 20%. After the deposition of the FEBID piezoresistors the structure resonance frequencies decreases again by ca. 15 kHz. As the frequency decrease is much smaller than after the FIB processing, it can be assumed that the frequency change correlates with the structure mass loading caused by the FEBID resistors. All the resonance curves presented in Fig. 12.14 were normalized in order to present a family of resonance characteristics of the same height.

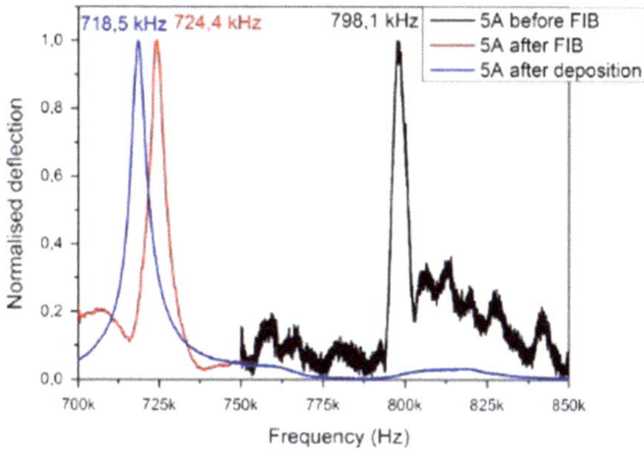

Fig. 12.14 Resonance curves of the silicon nitride bridge before and after FIB processing and after FEBID deposition of FEBID piezoresistors

The resonance properties were measured using an auxiliary piezoelectrical actuator biased with an high-resolution signal generator which excited vibration of the investigated structure.

The reference AFM silicon probes are used in the static deflection measurements of the silicon nitride bridges and the FEBID piezoresistors. In contrast to the standard AFM cantilevers, in the AdvancedTEC™ sensors the probe protrudes outside the spring beam. In this way the microprobe can be placed on the investigated bridge with high precision enabled by the optics integrated with the atomic force microscope. Moreover, in order to establish the relationship between photodiode and piezoelectric actuator displacement signals the cantilever deflection detector sensitivity was determined, when the cantilever was pushed against hard surface as it was described in [36]. In this way the so called load force (LF) curves can be recorded in the contact mode (CM) atomic force microscopy (AFM) technology and used for the precise and quantitative structure characterization. The AFM machine was integrated with the measurement equipment (current to voltage (I/U) converter Keithley 428, multimeter Agilent 2001) making it possible to record the changes of the FEBID piezoresistors resistance change under the load of the reference cantilever.

In Fig. 12.15 the measurement architecture is presented. The FEBID piezoresistors are biased using an integrated voltage source, the voltage response of the I/V converter is measured by the multimeter and recorded by the host PC computer. The proposed measurement architecture offers the highest possible resolution of the resistance change detection. In Fig. 12.16 the result of the performed investigation is shown. The microbridge was loaded with the reference cantilever [36]. The microstructure was deflected by 1 μm in every step. The resistance changes were approximated in order to calculate *GF*, which is the measure of the sensitivity of the deflection detection.

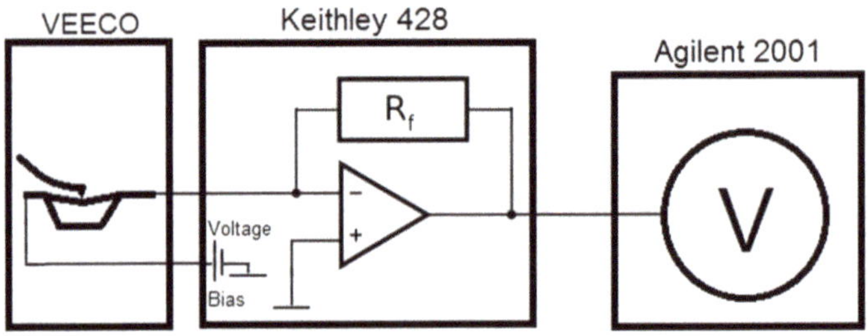

Fig. 12.15 Measurement setup for the characterization of the FEBID piezoresistors

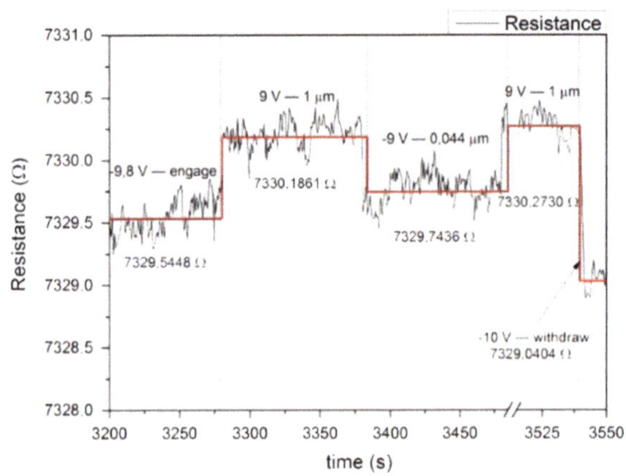

Fig. 12.16 Measurement of the FEBID resistance under microstructure loading/test deflection

The elongation of the bridge can be estimated according to the scheme:

(a) longitudinal elongation of the bridge under the load force:

$$L_{end} = 2 \times \sqrt{\left(\frac{1}{2}L_0\right)^2 + d^2}, \tag{12.23}$$

L_{end}—is the length of the microbridge after structure loading, L_0—the length before loading, d-the test loading

(b) gauge factor GF:

$$GF = \frac{\frac{\Delta R}{R}}{\frac{\Delta L}{L_0}}, \tag{12.24}$$

$$\Delta L = L_{end} - L_0, \tag{12.25}$$

$$\Delta R = R_{end} - R_0, \tag{12.26}$$

where R_{end}—is the resistance after loading, R_0—is the resistance before loading.

Basing on the above described calculations the *GF* of the platinum/carbon FEBID piezoresistors varies in the range from 4 to 6. The obtained result indicates that the measurement sensitivity is bigger than in the case of the bulk metal piezoresistors, when the *GF* varies in the range from 2 to 4. In addition, the performed experiments clearly show, that the proposed technology makes it possibly to fabricate and deposit strain sensors on the NEMS devices, at the positions where the highest stress occurs. The resolution, with which such a deposition can be done is of tens of nanometers indicating, that the described process is very attractive for the nanotechnological applications. Usage of the AFM based techniques makes it possible to determine the basic parameter of the developed devices.

12.4 Summary

NEMS technology opens new fields in nanotechnology. The extremely small dimensions of the fabricated devices make it possible to define the place of the investigations and measurement with the highest possible resolution. In this way investigations of the quantum phenomena will be enabled in many research laboratories active on this area. The growing range of NEMS applications is correlated with the need for device characterization and measurements. This symmetrical relation stimulates the progress in the nanotechnology and micro- and nanofabrication.

Acknowledgements This work was supported by the Wrocław University of Science and Technology (WUST) statutory grant. The author would like to thank all the coworkers of the Nanometrology Division of the Faculty of Microsystems Electronics and Photonics at the WUST for their support and collaboration.

References

1. Bannon F, Clark J, Nguyen C (2000) High-Q HF microelectromechanical filters. IEEE J Solid-State Circuits 35(4):512–526
2. Barton RA, Ilic B, Van Der Zande AM, Whitney WS, McEuen PL, Parpia JM, Craighead HG (2011) High, size-dependent quality factor in an array of graphene mechanical resonators. Nano Lett 11(3):1232–1236
3. Belic D, Shawrav M, Gavagnin M, Stöger-Pollach M, Wanzenboeck D, Bertagnolli E (2015) Direct-write deposition and focused-electron-beam-induced purification of gold nanostructures. ACS Appl Mater Interfaces 7(4):2467–2479
4. Chen C, Hone J (2013) Graphene nanoelectromechanical systems. Proc IEEE 101(7):1766–1779
5. Chen C, Rosenblatt S, Bolotin KI, Kalb W, Kim P, Kymissis I, Hone J (2009) Performance of monolayer graphene nanomechanical resonators with electrical readout. Nat Nanotechnol 4(12):861–867
6. Cleland A, Roukes M (1996) Fabrication of high frequency nanometer scale mechanical resonators from bulk Si crystals. Appl Phys Lett 69:2653
7. Cleland A, Roukes M (1999) External control of dissipation in a nanometer-scale radiofrequency mechanical resonator. Sens Actuators A 72(3):256–261

8. Goniszewski S, Gallop J, Adabi M, Gajewski K, Shaforost O, Klein N, Hao L (2015) Self-supporting graphene films and their applications. IET Circuits Devices Syst Spec 9:420–427
9. Gotszalk T, Grabiec P, Rangelow I (2003) Calibration and examination of piezoresistive Wheatstone bridge cantilevers for scanning probe microscopy. Ultramicroscopy 97(1–4):385–389
10. Grabiec P, Gotszalk T, Radojewski J, Edinger K, Abedinov N, Rangelow IW (2002) SNOM/AFM microprobe integrated with piezoresistive cantilever beam for multifunctional surface analysis. Microelectron Eng 61–62:981–986
11. Hoeflich K, Jurczyk J, Zhang Y, Puydinger M, Goetz M, Guerra-Nunez C, Best J, Kapusta Cz, Utke I (2017) Direct electron beam writing of silver-based nanostructures. ACS Appl Mater Interfaces 9:24071–24077
12. Huang S, Stott A, Green R, Beck M (1988) Electronic transducer for measurement of low value capacitances. J Phys E: Sci Instrum 21:242
13. Huang X, Zorman C, Mehregany M, Roukes M (2003) Nanoelectromechanical systems: nanodevice motion at microwave frequencies. Nature 421:6922
14. Huth M (2010) Granular metals: from electronic correlations to strain-sensing applications. J Appl Phys 107:113709
15. Ko WH (2007) Trends and frontiers of MEMS. Sens Actuators A 136(1):62–67
16. Koops H, Fukuda H (2016) Giant current density via indirect exciton orbit overlapping in polarized nano-granular materials. J Vac Sci Technol 33(2):02B108
17. Lewis B, Mound B, Srijanto B, Fowlkes J, Pharr G, Rack P (2017) Growth and nanomechanical characterization of nanoscale 3D architectures grown via focused electron beam induced deposition. Nanoscale 9:16349–16356
18. Li M, Tang HX, Roukes ML (2007) Ultra-sensitive NEMS-based cantilevers for sensing, scanned probe and very high-frequency applications. Nat Nanotechnol 2(2):114–120
19. Llobet J, Gerboles M, Sansa M, Bausells J, Borrise X, Perez-Murano F (2015) Fabrication of functional electromechanical nanowire resonators by focused ion beam implantation. J Micro-Nanolithography MEMS and MOEMS 14(3)
20. Llobet J, Sansa M, Gerbolés M, Mestres N, Arbiol J, Borrisé X, Pérez-Murano F (2014) Enabling electromechanical transduction in silicon nanowire mechanical resonators fabricated by focused ion beam implantation. Nanotechnology 25:135302
21. López-Polín G, Gómez-Navarro C, Parente V, Katsnelson MI, Pérez-Murano F, Gómez-Herrero J (2015) Increasing the elastic modulus of graphene by controlled defect creation. Nat Phys 11:26
22. Moczala M, Babij M, Kwoka K, Piasecki T, Sierakowski A, Gotszalk T (2019) Resolution improvement in electromagnetically actuated Wheatstone bridge configuration micromechanical resonators. Sens Actuators A 284:181–185
23. Moczała M, Kopiec D, Sierakowski A, Dobrowolski R, Grabiec P, Gotszalk T (2014) Investigations of mechanical properties of microfabricated resonators using atomic force microscopy related techniques. Microelectron Eng 119:164–168
24. Moczała M, Kwoka K, Piasecki T, Kunicki P, Sierakowski A, Gotszalk T (2017) Fabrication and characterization of micromechanical bridges with strain sensors deposited using focused electron beam induced technology. Microelectron Eng 176:111–115
25. Moczała M, Sierakowski A, Dobrowolski R, Grabiec P, Gotszalk T (2013) Fabrication and measurement of micromechanical bridge structures for mass change detection. Proceedings SPIE, vol 8902, p 89021.s
26. Nieradka K, Kopiec D, Małozięć G, Kowalska Z, Grabiec P, Janus P, Gotszalk T (2012) Fabrication and characterization of electromagnetically actuated microcantilevers for biochemical sensing, parallel AFM and nanomanipulation. Microelectron Eng 98:676–679
27. Orłowska K, Słupski P, Świątkowski M, Kunicki P, Sankowska A, Gotszalk T (2015) Light intensity fibre optic sensor for MEMS displacement and vibration metrology. Opt Laser Technol 65:159–163
28. Orłowska K, Światkowski M, Kunicki P, Kopiec D, Gotszalk T (2016) High-resolution and wide-bandwidth light intensity fiber optic displacement sensor for MEMS metrology. Appl Opt 55(22):5960–5966

29. Polski Komitet Normalizacyjny (2010) Międzynarodowy słownik metrologii. Pojęcia podstawowe i ogólne oraz terminy z nimi związane (VIM). PKN-ISO/IEC Guide 99
30. Puydinger M, Velo M, Domingos R, Zhang Y, Maeder X, Guerra-nun C, Be F (2016) Annealing-based electrical tuning of cobalt–carbon deposits grown by focused-electron-beam-induced deposition. ACS Appl Mater Interfaces 8:32496–32503
31. Rangelow IW, Grabiec P, Gotszalk T, Edinger K (2002) Piezoresistive SXM sensors. Surf Interface Anal 33:59–64
32. Schwalb Ch, Grimm Ch, Baranowski M, Sachser R, Porrati F, Reith H, Das P, Müller J, Völklein F, Kaya A, Huth M (2010) A tunable strain sensor using nanogranular metals. Sensors 10:9847–9856
33. Sekaric L, Parpia JM, Craighead H, Feygelson T, Houston B, Butler J (2002) Nanomechanical resonant structures in nanocrystalline diamond. Appl Phys Lett 81:4455–4457
34. Smith D, Pratt J, Howard L (2009) A fiber-optic sinterferometer with subpicometer resolution for dc and low-frequency displacement measurement. Rev Sci Instrum 80(3):035105
35. Swiatkowski M, Wojtuś A, Wielgoszewski G, Rudek M, Piasecki T, Jozwiak G, Gotszalk T (2019) A low-noise measurement system for scanning thermal microscopy resistive nanoprobes based on a transformer ratio-arm bridge. Meas Sci Technol 29:045901
36. Tamayo J (2005) Study of the noise of micromechanical oscillators under quality factor enhancement via driving force control. J Appl Phys 97(4):1–10
37. Tortonese M, Barrett R, Quate C (1993) Atomic resolution with an atomic force microscope using piezoresistive detection. Appl Phys Lett 62(8):834–836
38. Zaborowski M, Dumania P, Tomaszewski D, Czupryniak J, Ossowski T (2012) Development of Si nanowire chemical sensors. Proc Eng 47(1000):1053–1056. https://doi.org/10.1016/j.proeng.2012.09.331

Chapter 13
Numerical Case Studies: Forward Problems

Paolo Di Barba and Maria Evelina Mognaschi

13.1 Introduction

Electrostatic micromotors were the first MEMS which had been designed and prototyped exploiting Silicon integrated technology. This achievement took place at the Berkeley university laboratories first [14, pp. 41–47, 15, pp. 49–55], and at the MIT laboratories next [1, 23] in the period from late eighties through early nineties of last century. The cultural impact was great, because the miniaturization of a motor—well-known at the macro-scale—was implemented as a manufacturable process at the micro-scale. However, the technological success of micromotors was modest, because static friction soon turned out to be the major obstacle against the rotor movement. Nevertheless, from the modeling viewpoint, the rotating micromotor is a good example of a simple device exhibiting the main features of MEMS, like e.g. mechanical and electrical properties of poly-silicon, size scale of components, and values of field strength. Therefore, it could be considered as a kind of computational benchmark useful to assess a numerical method for automated optimal design: some electrostatic micromotors are analyzed in Sects. 13.2 and 13.3.

Over the last decade, extensive efforts have been devoted to the development of micro-accelerometers for different applications such as automotive safety, navigation, audio-video and health monitoring. MEMS accelerometers have an excellent sensitivity and a wide dynamic range and at the same time they are low cost and can be mass produced. There are many different physical mechanisms which MEMS accelerometers can be based on e.g. piezoresistive, electromagnetic, piezoelectric, ferroelectric, capacitive and so on. The example proposed in Sect. 13.4, the comb drive micro-actuator, is focused on the capacitive effect. The comb drive device has the advantage of small size, weight, and excellent sensitivity [21].

Among different ways of exciting MEMS, the magnetic actuation is one way which exhibits many advantages. In fact, it allows to have a good linearity of movement versus the excitation signal, low voltages needed for power supply and hence low power consumption and finally they are simple to control [16, 17]. An example of magnetic MEMS is the magnetic micromirror actuator, shown in Sect. 13.5.

P. Di Barba and S. Wiak, *MEMS: Field Models and Optimal Design*, Lecture Notes in Electrical Engineering 573, https://doi.org/10.1007/978-3-030-21496-8_13

However, one of the most used applications in the field of MEMS is the electro-thermo-elastic actuator. It can show different shapes (e.g. two-arms or three-arms device) [9] but common characteristics are repeatable behaviour, long life-time [20], easy to produce because of its simple design and monolithic single-material structure [19]. The most appreciated characteristics of these MEMS are the large force densities and large displacements, with comparison to other MEMS actuated by e.g. electrostatic field. Two examples of electro-thermo-elastic actuators are shown in Sects. 13.6 and 13.7.

Another popular MEMS device is a singly clamped beam (known commonly as cantilever). Its most important applications involve atomic force microscopy technologies, in which a cantilever with an integrated tip scans a sample and the beam deflection corresponds with the interactions between the tip and the surface [22]. The cantilever-based sensing systems have become very attractive in microbiology and biotechnology: an example of electromagnetically actuated cantilever is described in Sect. 13.8.

To summarize, different numerical examples are shown in this Chapter, the most of which are analyzed by means of finite element models. In fact, due to MEMS nonlinearity and the presence of couplings of energy domains like electromagnetics and mechanics, multiphysics models, suitable to be solved by means of numerical method e.g. FEM, are needed [11].

13.2 Axial-Field Electrostatic Micromotor

A rotating micromotor with axial field has been considered (Neuchâtel prototype) [24]; it realizes the so-called top-drive device. Due to reduced stray field, the latter offers stronger variations of capacitance and so higher drive torque; on the other hand, applied voltage cannot be very high in order to avoid the bending effect of the rotor tooth towards the substrate. In Fig. 13.1 longitudinal and cross-sectional views of the micromotor are represented, respectively.

The device is characterized by $N_s = 12$ stator electrodes that are placed underneath $N_r = 8$ rotor teeth (3/2 machine). The rotor is supported by four hemispherical bushings and has a diameter equal to 200 μm, while the air-gap width is equal to 1.5 μm; electrodes and teeth have the same angular width equal to 20°; moreover, $R_{ir} = 30$ μm, $R_a = 50$ μm, $R_i = 60$ μm. Rotor and stator regions can be assimilated to ideal conducting media since they are realized in phosphorus-doped polycrystalline silicon. The device, acting as a position actuator, is supplied by a three-phase system of rectangular voltages $V = 50$ V.

In order to develop the circuit model of the device, a circuit approach has been adopted. The main assumptions made can be summarized in this way:

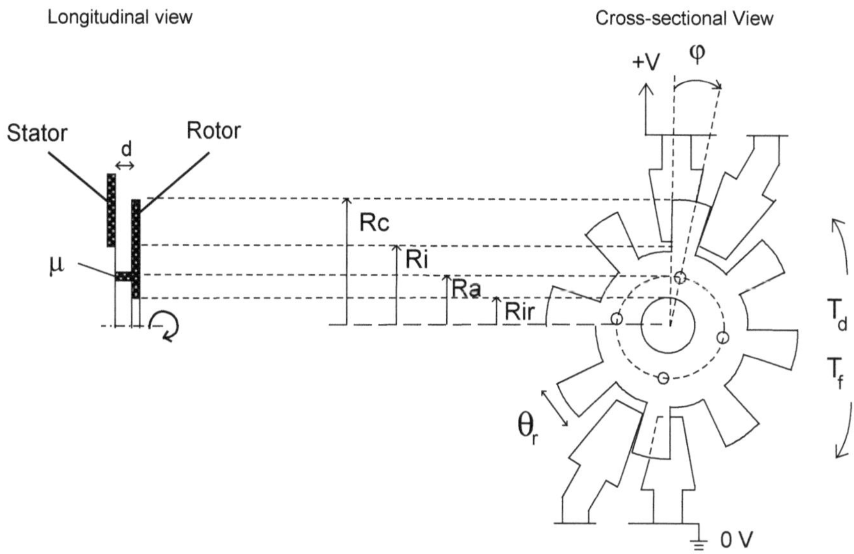

Fig. 13.1 Geometrical model of the top-drive micromotor

- the electric field takes place only in the axial air-gap, so that any fringing field effect is neglected;
- the variation of the electrostatic energy stored in the air-gap is a piecewise linear function of the rotor position;
- the device is in no-load condition.

Since the device operates at a constant voltage V, co-energy $W(\phi)$ is easily estimated by the parallel-plate capacitance as:

$$W = \frac{\varepsilon_0}{4d}\left(R_c^2 - R_i^2\right)V^2(\vartheta_r - |\varphi|) \tag{13.1}$$

Applying the principle of virtual work, the maximum drive torque M_d corresponding to the aligned position between rotor tooth and energized stator electrode is:

$$M_d = \frac{\varepsilon_0}{4d}\left(R_c^2 - R_i^2\right)V^2 \tag{13.2}$$

Following the same way, the maximum axial force F_a clamping the rotor on the stator plane is:

$$F_a = -\frac{\varepsilon_0}{4d^2}\left(R_c^2 - R_i^2\right)V^2\vartheta_r \tag{13.3}$$

Since it acts on the rotor bushings, the associated static friction torque is:

$$M_f = \mu R_a F_a \tag{13.4}$$

where $\mu = 0.2$ has been assumed as the coefficient of dry friction of poly-Si.

The model provides the maximum effective torque (i.e. the net torque) of the micromotor as:

$$M_e = |M_d| - |M_f| = \frac{\varepsilon_0}{4d}\left(R_c^2 - R_i^2\right)V^2\left(1 - \frac{\mu R_a \vartheta_r}{d}\right)\frac{N_s}{3} \tag{13.5}$$

It is assumed that only one phase is supplied at a time; therefore, $N_s/3$ capacitors are energized. In Fig. 13.2 the contributions of drive and friction to static torque are represented as functions of the air gap width d.

It can be noted that the amount of friction for the prototype ($d = 1.5$ μm) is higher than the drive term so that the motor cannot start. In Fig. 13.3 the behaviour of net torque versus air gap is represented.

When the air gap width is around 3.5 μm friction and drive terms are equal, so that net torque is zero. The optimal value of air gap width is $d = 6.98$ μm; correspondingly, net torque is maximum and equal to 10.132 pNm.

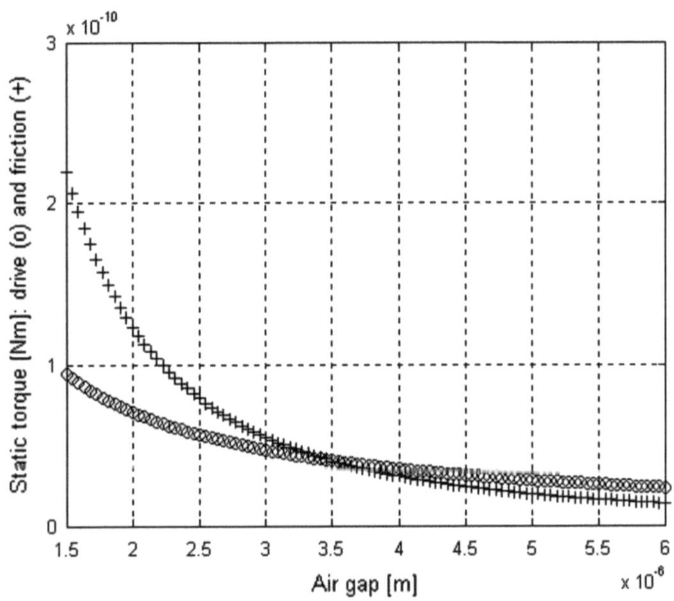

Fig. 13.2 Dependence of drive and friction torques on air gap width

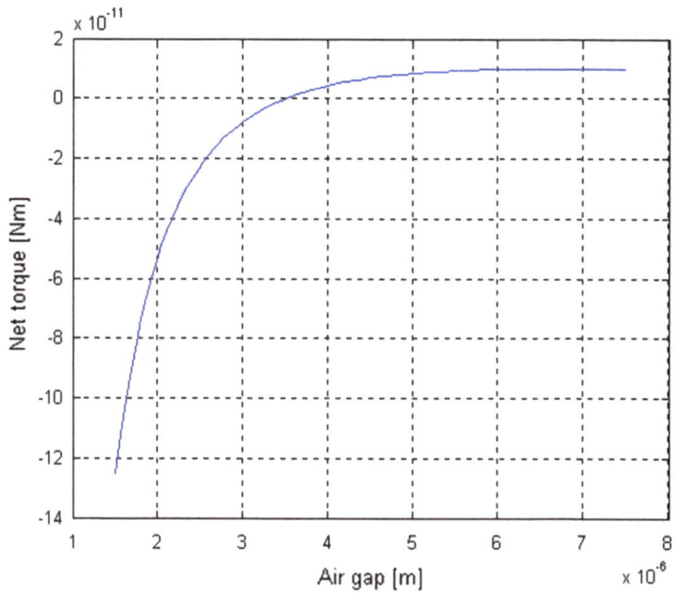

Fig. 13.3 Net torque as a function of air gap width

13.3 Radial-Field Electrostatic Micromotor

The device considered is a side-drive synchronous micromotor; it is characterized by a salient pole geometry exhibiting stator poles and rotor teeth. The rotor—supported by hemispherical bushings—is free to rotate around a bearing; in Fig. 13.4 the device is shown: the salient-pole geometry is evident.

The electrical supply provides a three-phase system of rectangular voltages of hundred volts: every third stator pole is electrically connected to the same phase.

13.3.1 Zero-Dimensional Model

The motor under study is a 12/8 motor and it has the following characteristics [10]: each stator and rotor poles of the motor has an angular width equal to 18°; the space between stator poles is 12° wide, while the angle between rotor teeth is equal to 27° (see Fig. 13.5). The rotor has radius equal to 50 μm; the gap between a rotor tooth and an aligned stator pole is 1.5 μm wide, the inner diameter of the rotor is equal to 21 μm, while the axial thickness of rotor and stator is of 2.2 μm.

For an approximate evaluation of the torques acting upon the micromotor, the use of a simple equivalent capacitive circuit is possible. There are two major issues in defining a lumped-parameter model: how to subdivide the continuum device into a

Fig. 13.4 Photo of the micromotor

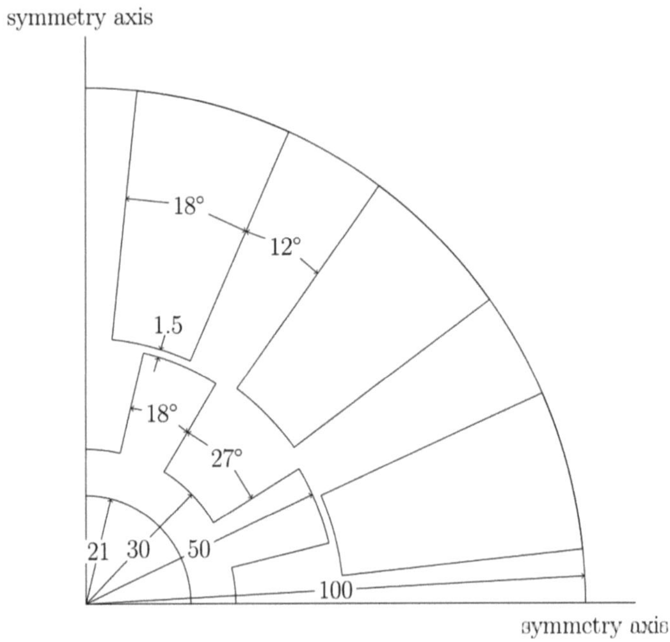

Fig. 13.5 Geometry of the electrostatic micromotor (measurement unit in μm)

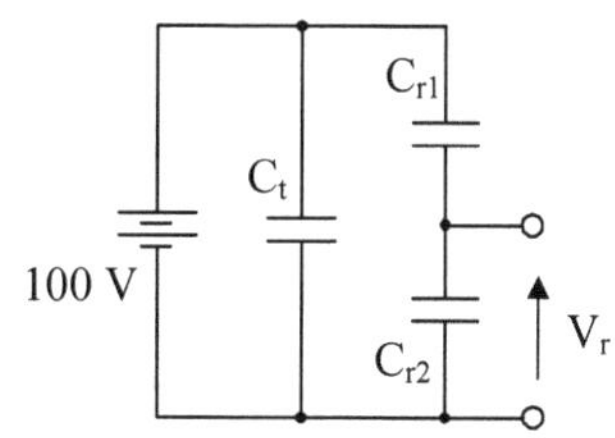

Fig. 13.6 1D equivalent circuit

network of capacitors and how to determine the capacitance values for each element. The problem of subdivision is particularly troublesome because, due to electric properties of materials involved, there is no univocal correspondence between geometry and network topology.

Moving from this background, the electric field is supposed to be concentrated within stator and rotor and so be radial (side-drive micromotor); furthermore, it is assumed that rotor and stator are made up of conducting material, so to consider the respective surfaces equipotential. By properly subdividing the air gap in cylindrical sectors, analytical expressions are obtained for the global capacitance at a position φ; the axial dimension of the equivalent capacitor is assigned the rotor thickness.

In Fig. 13.6, a simple equivalent circuit is sketched, suitable for the class of studied devices. Capacitor C_t has got constant capacitance, and represents the global effect of the tangential capacitances between the power supplied electrode and the nearby grounded ones. Capacitors C_{r1} and C_{r2} model the capacitive coupling of the electrode with the rotor and of the rotor with the ground, respectively; hence, their capacitance is variable with the angular position. Potential V_r, which varies with the angular position, is the rotor potential.

Given a circuit model, a preliminary and fast estimation of torques can be carried out. In particular, the static torque M is a quantity of primary concern for the device designer. The basic formula for computing it is

$$M(\varphi) = \frac{1}{2} V^2 \frac{\partial C_{eq}}{\partial \varphi} \tag{13.6}$$

where C_{eq} is the equivalent capacitance, φ is the rotation angle and V is the voltage supply.

By applying the circuit model, four equivalent capacitors are simultaneously energized by a voltage, equal e.g. to 100 V, while the other ones are grounded. As concerns the rotor movement, an angular step of 3° has been considered; in Fig. 13.7 the detail of two different stator-rotor positions is shown.

Six different relative positions, covering a pole pitch, have been considered to evaluate electrostatic energy; by applying formula (13.6), the results reported in Table 13.1 have been obtained. The static torque has been scaled by the actual thickness (2.2 pm) of the micromotor.

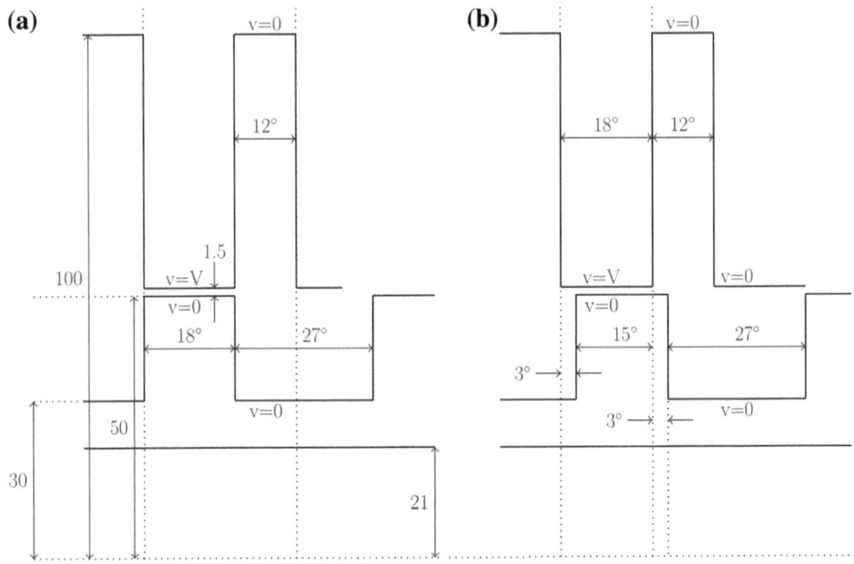

Fig. 13.7 Developed view of circuit model (linear dimensions in μm): **a** aligned pole-tooth position; **b** position displaced by 3°

Table 13.1 Results of circuit analysis (the position 0° is the aligned position, see Fig. 13.7a)

Displacement (°)	Total capacitance (pF)	Torque (pNm)
0	96.540	–
3	83.740	10.76
6	68.360	12.92
9	52.980	12.92
12	37.604	12.92
15	22.820	11.97

13.3.2 Two-Dimensional Model

The motor under study is a 18/6 one and it has been prototyped with the following data: inner and outer rotor radii 40 and 60 μm, stator radius 63 μm, angular width of rotor slot 40° [6, 8]. Accordingly, the width of the stator-to-rotor air-gap varies between 3 and 23 μm. The stator electrodes are supplied by a three-phase system of square voltages, equal to 100 V, while the rotor potential is floating.

In order to evaluate quantitatively the system, the field model of the device has to be adopted. After introducing a system of cylindrical coordinates (ρ, φ, z), in terms of scalar potential V the Laplace equation governing the electric field E can be expressed as:

$$\nabla \cdot (\varepsilon \nabla V) = 0 \ \ in \ \ \Omega \tag{13.7}$$

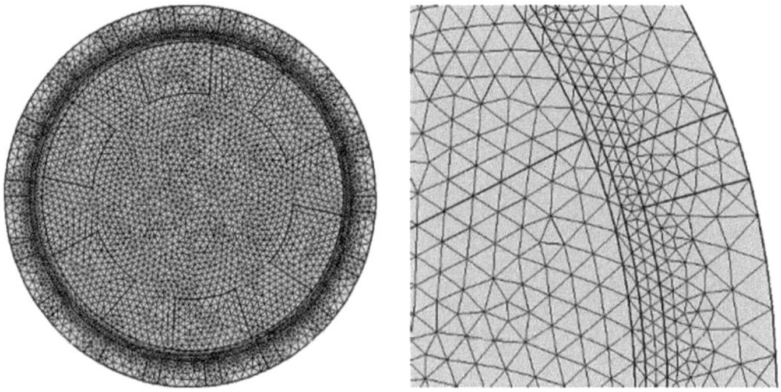

Fig. 13.8 Electrostatic micromotor: left—Finite-element mesh right—mesh detail

$$V = 0 \quad on \quad \Gamma_1 \tag{13.8}$$

$$V = V_0 \quad on \quad \Gamma_2 \tag{13.9}$$

$$\frac{\partial V}{\partial n} = 0 \quad on \quad \Gamma_3 \tag{13.10}$$

where ε is the permittivity and V_0 is the applied voltage. Knowing the scalar potential V, the electric field strength E can be evaluated as:

$$-\nabla V = \overline{E} \tag{13.11}$$

It can be noticed that only one phase is considered, so that the poles adjacent to the supplied one are grounded.

In order to solve the boundary value problem (13.7)–(13.11) numerically, the finite element method was used. A dense 2-D mesh was created, composed by about 9,000 elements with linear variation; it is shown in Fig. 13.8.

A typical solution of the field problem in shown in Fig. 13.9.

In order to evaluate the torque of the motor versus the angular position, different relative positions between stator and rotor are obtained by rotating the rotor subregion by one step $\Delta\varphi$. For calculating the force at each position, it is possible to apply either the virtual work principle or the Maxwell's stress tensor.

The first method implies to evaluate the co-energy W' of the system as:

$$W' = \frac{1}{2} \int_{\Omega} \varepsilon E^2 d\Omega \tag{13.12}$$

assembling all the elementary contributions over the physical domain Ω.

Following the procedure described, a set of energy-angle points is obtained and, after numerical interpolation, a continuous energy-angle curve is then carried out.

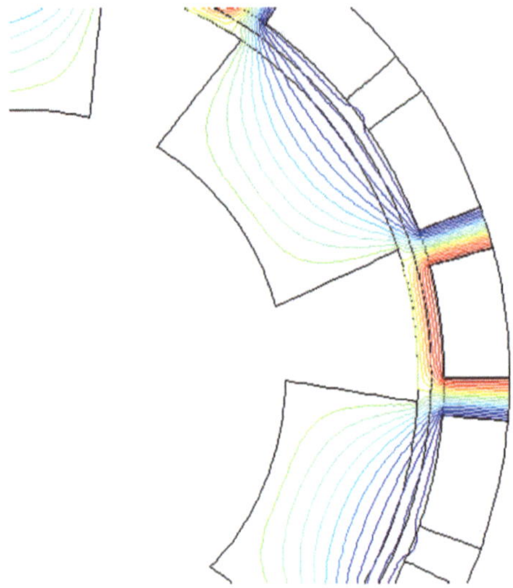

Fig. 13.9 Potential contour lines

Since the device considered operates at a rated value of supply voltage V_0, the torque M can be given, as the derivative of the co-energy W' stored in the field region with respect to the angular displacement φ of the rotor:

$$M(\varphi) = \frac{\partial W'}{\partial \varphi} \tag{13.13}$$

Applying the Maxwell's stress tensor means calculating the force F acting on each point P of the closed surface Γ, surrounding the volume Ω, as

$$\overline{F} = \int_{\Omega} \overline{\nabla} \cdot \overline{\overline{T}} \, d\Omega = \int_{\Gamma} \overline{\overline{T}} \cdot \bar{n} \, d\Gamma \tag{13.14}$$

where $\bar{n}$ is the outward normal unit vector and $\overline{\overline{T}}$ is the Maxwell's electric stress tensor, defined in Eq. (3.69).

Correspondingly, the torque is given by

$$\overline{M} = \int_{\Gamma} \bar{r}_{PO} \times \overline{\overline{T}} \cdot \bar{n} d\Gamma \tag{13.15}$$

where $\bar{r}_{PO}$ is the position vector of point P on Γ with respect to the rotation axis in O.

In this 2D example, Γ is a closed line, in particular it is the circumference laying in the middle of the airgap, which encloses the region of the rotor.

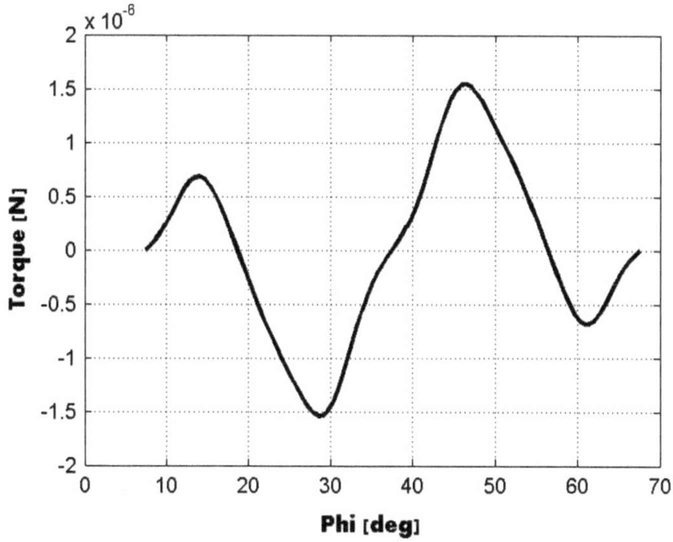

Fig. 13.10 Electrostatic micromotor: torque-angle curve on a pole pitch

The closed line over which the integral (13.15) is calculated must enclose entirely the part of device which torque or force has to be calculated and, at the same time, must not intersect it.

Moreover, it can be demonstrated that, if the abovementioned conditions are fulfilled, the length or the shape of the closed line has no influence on the final result.

The static torque calculated for the prototype, considering an axial length of 1 m i.e. the torque per meter length (N) is shown in Fig. 13.10.

For both methods, the refinement of the airgap is the most critical; in order to have the fewest modifications to the mesh while rotating the rotor from one position to the next, it is recommended that the mesh nodes along the so-called "slipping line" i.e. a circle line passing through the middle of the air-gap, are set regularly spaced by the angular distance $\Delta\varphi$.

Moreover, in order to extract a smooth electric field at the airgap and a precise evaluation of the electric Maxwell's stress tensor, it is mandatory that the airgap mesh is fine enough.

13.4 Comb Drive Accelerometer

The comb drive accelerometer consists of two finger structures, called fixed finger and movable finger. The fixed fingers are fixed to the frame. The movable fingers are attached to a proof-mass and suspended by flexible elements, e.g. springs, to the frame. The device senses any external acceleration which is transferred to the proof-mass through the flexible elements. The movement of the movable structure changes the capacitance between the fixed finger and the movable finger [25].

A prototype model of the comb drive device [27], which cross-section is shown in Fig. 13.11, where s.a. means the symmetry axis, is here analyzed. It is characterized by the following parameters: $w_m = 4\ \mu m$, $w_f = 4\ \mu m$, $h_m = 2\ \mu m$ and $h_f = 2\ \mu m$; hence w and h are width and height of the movable (m) and the fixed (f) electrodes, respectively.

The device exhibits $10 + 9$ electrodes. The geometry of the device is amenable to the lumped-parameter model proposed in Tang et al. [26]. Moreover, the fixed and movable electrodes of the comb drive are $2\ \mu m$ thick and $4\ \mu m$ wide, respectively; the air-gap distance g between them is $2\ \mu m$ wide, the same as their distance $z_1 = z_2$ from the grounded substrate.

The equation governing the analysis problem of the modeled device is the Laplace's equation of the electric scalar potential V in the computational domain (Eq. 13.7). Second-order Lagrangian shape functions were considered in the finite-element model: a typical mesh is composed of 170,000 elements with 240,000 unknowns [2, 3, 13], approximately, and it is shown in Fig. 13.12.

The device is considered electrically isolated: the boundary condition of the air subdomain is set to zero charge density; moreover, the fixed electrodes are at the

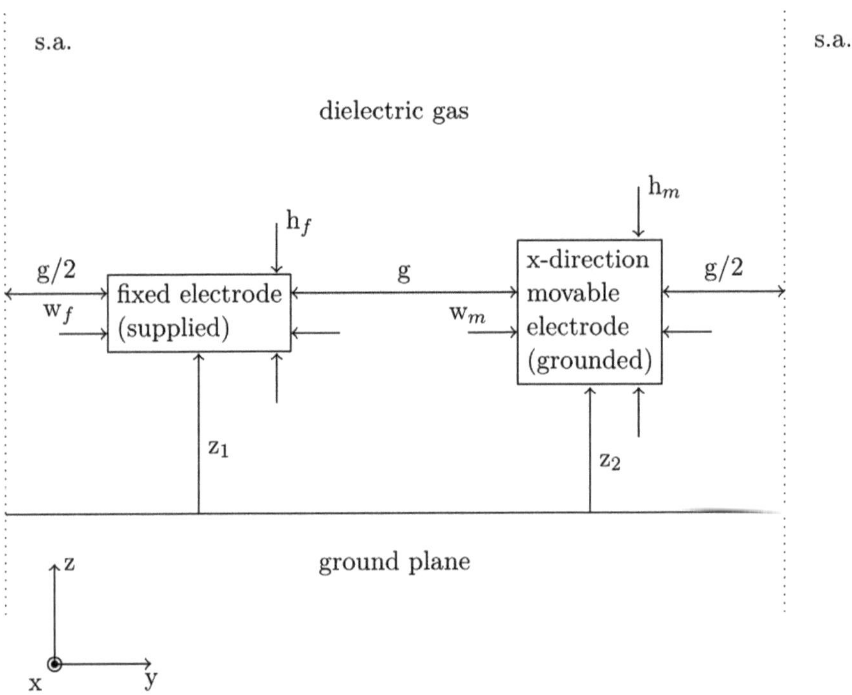

Fig. 13.11 Cross-section of the comb drive

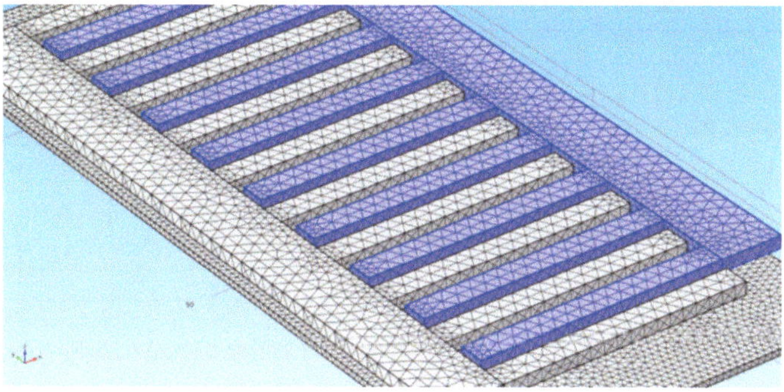

Fig. 13.12 Comb drive mesh

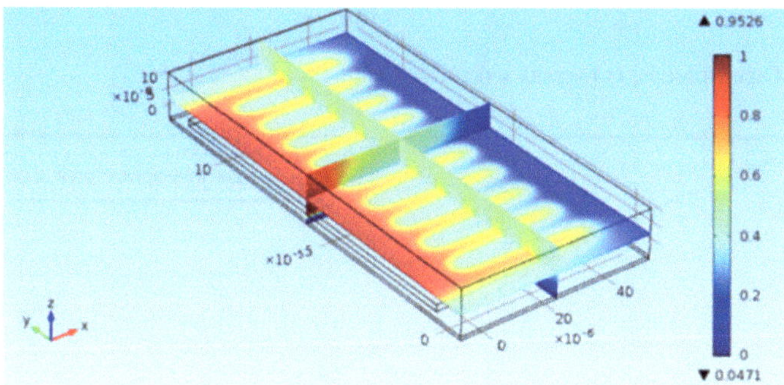

Fig. 13.13 Potential contour lines

same potential as the grounded substrate, while the movable electrodes are subject to voltage $V_0 = 1$ V. The device components are made of polycrystalline Si exhibiting a relative permittivity $\varepsilon_r = 4.5$.

The distribution of electric potential V is shown in Fig. 13.13.

In addition to the drive force in the direction of electrodes (shortly, x-directed drive force), the force due to electric field in the orthogonal direction (shortly, z-directed levitation force) takes place. The two force–displacement curves (drive and levitation, respectively) have been computed by means of the Maxwell's stress tensor method, taking the surface of the grounded electrodes as the integration surface. The elementary displacement was equal to 1 μm in the x-direction (14 steps) and 0.3 μm in the z-direction (7 steps): results are shown in Figs. 13.14 and 13.15, respectively.

The calculated force values are in agreement with reference values [26]; in particular, the approximated model $F_z = k(z - z_0)$ holds for the levitation term: k is named "electrostatic spring" constant, i.e., the z-directed force density per unit voltage (N V^{-2} m^{-1}), while z_0 is the equilibrium height of the movable electrode in the absence of a return spring force, i.e., the height toward which the electrodes spontaneously tend to move. It can be remarked that the drive force abruptly decreases for small displacements between fixed and movable electrodes (see Fig. 13.14), while

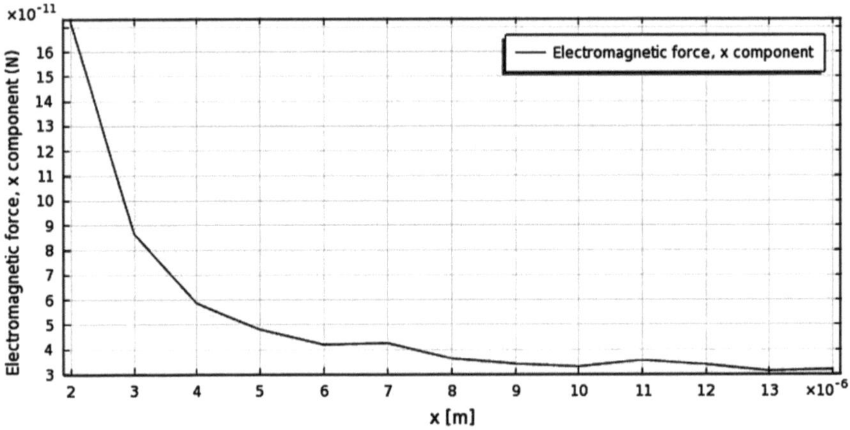

Fig. 13.14 Drive force F_x (x) versus x-displacement

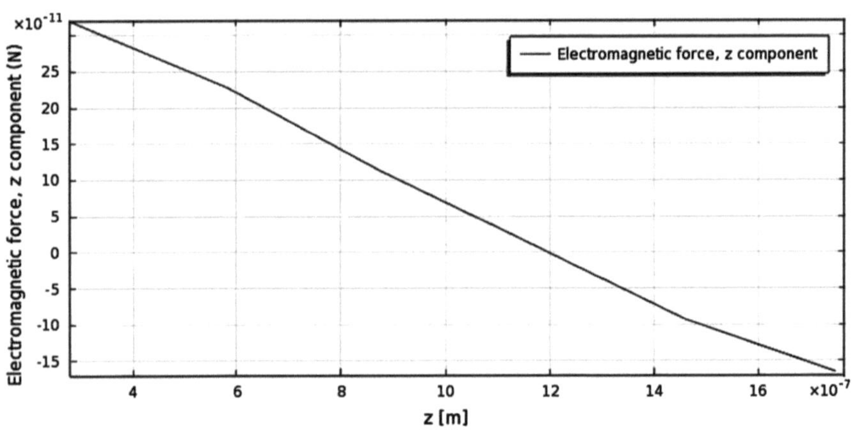

Fig. 13.15 Levitation force F_z (z) versus z-displacement

for larger displacements it tends to be constant. Moreover, according to the levitation force versus angle curve, the equilibrium point z_0 is located at 1.2 μm with respect to the substrate (see Fig. 13.15).

13.5 Magnetic MEMS: Micromirror Actuator

A micromagnetic device used as an optical switch [4] is considered. It is based on the rocking actuation of a ferromagnetic plate under the influence of a magnetic field variation. It consists of an NdFeB magnet, two conductors carrying like currents, and a ferromagnetic plate free to rotate around its axis, as shown in Fig. 13.16.

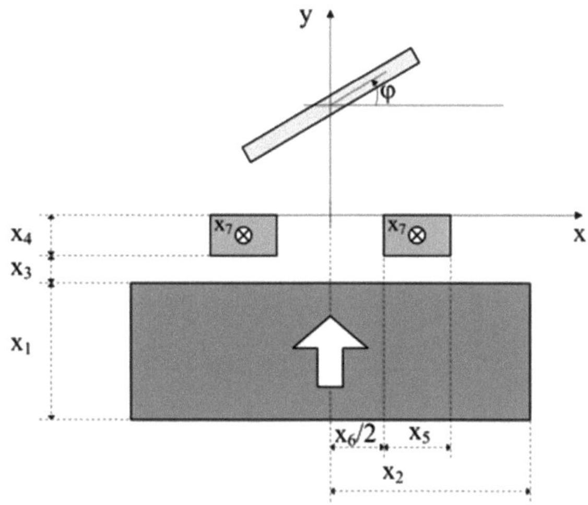

Fig. 13.16 Geometry of the micromirror

The residual induction of magnet is 1.2 T, while values of relative magnetic permeability equal to 10^3 and 1.05 are assumed for plate and magnet, respectively; moreover, the depth of the model is $w = 500$ μm. Referring to Fig. 13.9, the device has: $x_1 = 100$ μm, $x_2 = 1.2$ mm, $x_3 = 50$ μm, $x_4 = 50$ μm, $x_5 = 200$ μm, $x_6 =$ 600 μm, $x_7 = 10$ A, plate length = 1 mm, and plate height = 25 μm [12, 13].

The torque holding the plate at the prescribed angle is due to the field of the permanent magnet in the absence of current, while the actuation torque necessary to switch the plate angle is due to the field variation caused by a current pulse in the conductors. The field analysis problem consists of finding the magnetic field distribution for a given plate angle. The problem to be solved, in terms of magnetic vector potential $\overline{A}$ (see Sect. 4.3), is:

$$\nabla^2 \overline{A} = -\mu \overline{J} - \nabla \times \overline{B}_0 \tag{13.16}$$

where $\overline{J}$ is the coil current, μ is the magnetic permeability and $\overline{B}_0$ is the magnetic remanence of the permanent magnet. The magnetic induction field $\overline{B}$ is then calculated as

$$\overline{B} = \nabla \times \overline{A}. \tag{13.17}$$

In this respect, a typical finite-element mesh is composed of 2,300 triangles; second-order polynomial Lagrangian elements are considered, originating approximately 75,000 unknowns. The corresponding torque–angle curve (see Fig. 13.17) has been computed based on the virtual work principle.

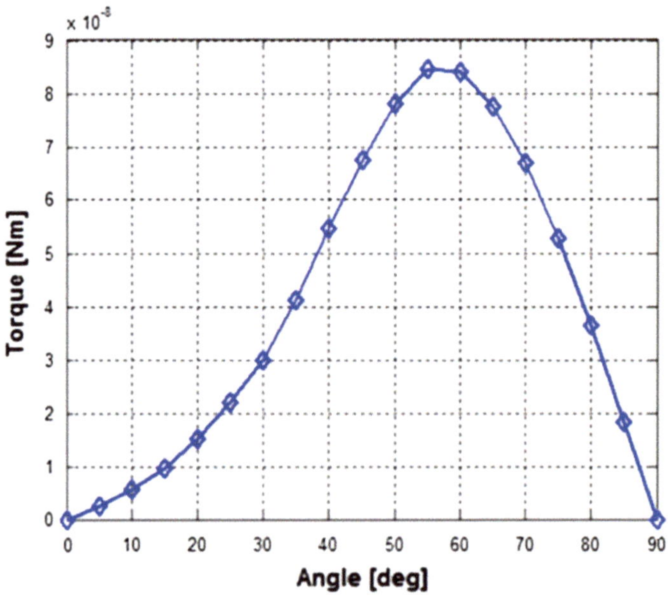

Fig. 13.17 Torque versus rotation angle

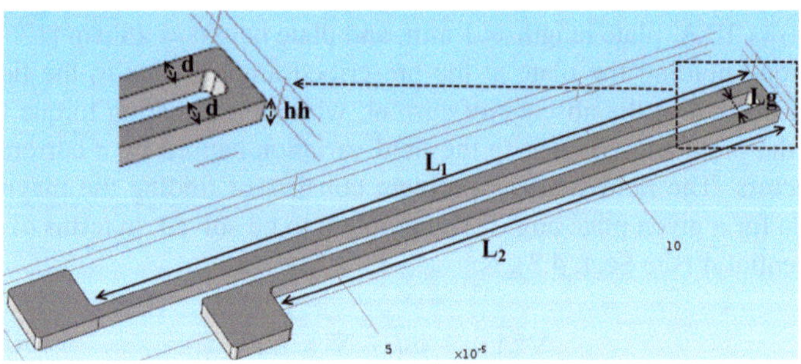

Fig. 13.18 Geometry of the two-arms microactuator

13.6 Two-Arms Electro-Thermo-Elastic Microactuator

The laterally-driven polysilicon microactuator considered as a case study [9] is the one studied in [18], where an analytical model of displacement and temperature is calculated. The microactuator is composed of two arms exhibiting different length (see Fig. 13.18); due to an applied voltage, current flows in the series-connected arms. The current heats the device and, because of the difference in length, the two heated arms elongate differently.

The beam is composed of two polysilicon layers: the top layer (thickness h_c) is conductive and the bottom layer (thickness h_u) is nonconductive. Under steady-state conditions, the top and bottom layers are assumed to have the same temperature.

The deflection u of the actuator depends on the following geometrical parameters: the length of the arms L_1 and L_2, the length L_g of the gap between long and short arm, the height h_h of the device and the width d of the arms. In particular, one has:

$$u = \frac{L_2^2}{6EI}(3X_3 - X_1 L_2) \tag{13.18}$$

where X_1 and X_3 come from the solution of the following system of equations (force method):

$$\begin{bmatrix} f_{11}\, f_{12}\, f_{13} \\ f_{21}\, f_{22}\, f_{23} \\ f_{31}\, f_{32}\, f_{33} \end{bmatrix} \begin{bmatrix} X_1 \\ X_2 \\ X_3 \end{bmatrix} = \begin{bmatrix} \Delta L_g \\ \Delta L_2 - \Delta L_1 \\ 0 \end{bmatrix}. \tag{13.19}$$

where X_1, X_2 and X_3 are the three redundants and, in particular, X_1 stands for a unit horizontal force, X_2 a unit vertical force and X_3 a unit couple force [18]. The terms f_{ij} represent the flexibility coefficients, given by:

$$f_{11} = \frac{L_2^2}{EI}\left(\frac{L_2}{3} + L_g + \frac{L_1}{3}\right) \tag{13.20}$$

$$f_{12} = f_{21} = -\frac{L_g L_2}{2EI}(L_g + L_1) \tag{13.21}$$

$$f_{13} = f_{31} = -\frac{L_2}{EI}\left(\frac{L_2}{2} + L_g + \frac{L_1}{2}\right) \tag{13.22}$$

$$f_{22} = \frac{L_g^2}{EI}\left(\frac{L_g}{3} + L_1\right) \tag{13.23}$$

$$f_{23} = f_{32} = \frac{L_g}{EI}\left(\frac{L_g}{2} + L_1\right) \tag{13.24}$$

$$f_{33} = \frac{L}{EI} \tag{13.25}$$

where E is the Young's modulus of polysilicon and $I = h_h d^3/12$ is the momentum of inertia.

In turn, the maximum temperature depends, from the geometrical viewpoint, on height h_h, being $h_u = h_h - h_c$:

$$T = \frac{V_0^2}{8k\rho}\left(\frac{1}{1 + \xi h_u / h_c}\right) + T_s \tag{13.26}$$

Table 13.2 Material properties of the microactuator

Parameter	Value	Parameter	Value
Thickness of conducting layer h_c (μm)	0.5	Thermal expansion coefficient α (°C^{-1})	2.7×10^{-6}
Thickness of top layer h_u (μm)	$h_h - h_c$	Young modulus E (Pa)	150×10^9
Parameter ξ	1/9	Substrate temperature T_s (°C)	20
Thermal conductivity k (W m^{-1} K^{-1})	41	Electrical resistivity ρ (Ω m)	5×10^{-4}

The parameter values used in this example are shown in Table 13.2.

The parameter ξ is used to take into account the difference of thermal conductivity and non-uniform heat generation between the top and the bottom layer [18]. The applied voltage is 5 V.

13.7 Three-Arms Electro-Thermo-Elastic Microactuator

An electro-thermo-elastic microactuator, characterized by 3D geometric model and coupled field analysis, is considered. The device, shown in Fig. 13.19, has length L in the range of hundreds of microns and width h_h in the range of tens of microns. An electric voltage is applied between two electrodes A and B; therefore, an electric current I flows in two out of three arms of the device (hot arms), while the third arm is current free (cold arm).

The actuator is fixed to a substrate at the three arm ends, while three cylindrical bushings act on the cold arm, in order to make it rotate in the xy-plane (Fig. 13.19). The deformation due to the overheating of the hot arms with respect the cold one is responsible for the rotation of the actuator.

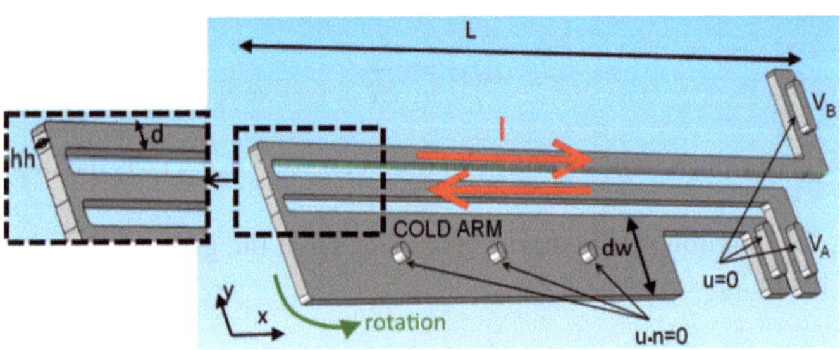

Fig. 13.19 Geometry of the microactuator (electrical and mechanical boundary conditions are also shown)

Table 13.3 Material properties of the microactuator

Material property	Value	Material property	Value
Electric conductivity σ ($S\ m^{-1}$)	5×10^4	Young modulus (Pa)	210×10^9
Thermal conductivity k ($W\ m^{-1}\ K^{-1}$)	34	Fracture toughness (Pa)	$1.44–2.51 \times 10^9$
Thermal expansion coefficient α (K^{-1})	2.6×10^{-6}	Poisson coefficient	0.22

A 3D finite-element model of the device has been developed [5, 9]; a typical finite-element mesh is composed of about 8,000 tetrahedra. The material properties used in the model are listed in Table 13.3.

The following multiphysics equations, which are coupled at the right-hand side level, are subsequently solved.

13.7.1 Electric Problem

The governing equations are

$$\nabla \cdot \sigma \nabla V = 0 \tag{13.27}$$

and

$$\overline{J} = \sigma \overline{E} = -\sigma \nabla V \tag{13.28}$$

subject to boundary conditions

$$V_A = const, \quad V_B = 0, \quad \overline{n} \cdot \overline{J} = 0 \ elsewhere \tag{13.29}$$

where $\overline{J}$ is the current density, $\overline{E}$ the electric field and V the unknown electric voltage.

13.7.2 Thermal Problem

The governing equation is

$$-\nabla \cdot (k \nabla T) = \overline{J} \cdot \overline{E} \tag{13.30}$$

subject to boundary conditions

$$T = 293.15 \text{ K } \textit{at the bushings and at the hinges} \tag{13.31}$$

$$-\bar{n} \cdot (k\nabla T) = h(T_{ext} - T) \ \textit{elsewhere} \tag{13.32}$$

with $\overline{J} \cdot \overline{E}$ the specific power due to the Joule effect, T the unknown temperature and h the convection coefficient; the latter has different values on the upper and the lower surface. In particular, the value of h is equal to 400 W m^{-2} K^{-1} for the upper surface and 2×10^4 W m^{-2} K^{-1} for the lower surface. These values are obtained by dividing the thermal conductivity of air (0.04 W m^{-1} K^{-1}) by the distance to the surrounding surfaces for the system.

13.7.3 Structural Problem

In order to solve the structural problem, the total strain ε is additively decomposed into its elastic $\overline{\varepsilon}^e$ and thermal $\overline{\varepsilon}^T$ components

$$\overline{\varepsilon} = \overline{\varepsilon}^e + \overline{\varepsilon}^T \tag{13.33}$$

The thermal component, which reads

$$\overline{\varepsilon}^T = \alpha(T - T_{ref}) \tag{13.34}$$

is known after the cascade solution of (13.27) and (13.32).

This way, the classical continuum-mechanics problem, that incorporates compatibility, equilibrium and elastic equations, can be cast; they read

$$\overline{\varepsilon} = \frac{1}{2}\left(\nabla\overline{u} + \nabla\overline{u}^T + \nabla\overline{u}\nabla\overline{u}^T\right) \tag{13.35}$$

$$\nabla \cdot \overline{S} = 0 \tag{13.36}$$

$$\overline{\varepsilon}^e = \overline{C}^{-1}\overline{S} \tag{13.37}$$

respectively, where u is the unknown displacement subject to

$$u_z = 0 \ \textit{at the bushings}, \quad u = 0 \ \textit{at the hinges} \tag{13.38}$$

while $\overline{S}$ is the second Piola-Kirchhoff stress tensor, and $\overline{C}$ is the elastic constitutive tensor.

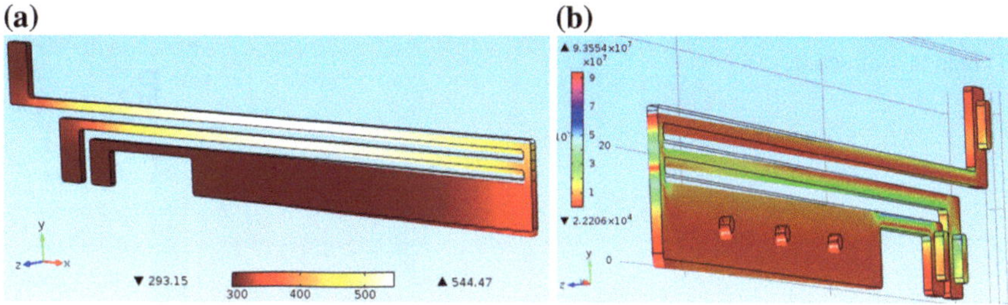

Fig. 13.20 Temperature (**a**) and stress (**b**) field maps

From the numerical viewpoint, the structural problem (13.33)–(13.38) is solved in the large displacement regime using the second Piola-Kirchhoff tensor as the stress measure, then the finite-element mesh is updated to simulate the deformed structure.

In Fig. 13.20 a solution of the multiphysics direct problem is shown; in particular, the field maps of temperature and stress are shown, respectively.

13.8 Electromagnetically Actuated Cantilever

The cantilever under study is a silicon cantilever in which a copper path is embedded. Its dimensions are length = 500 μm, width = 100 μm, height = 10 μm, conductor width = 10 μm.

Electromagnetic actuation of the cantilevers is a method for the movement control. The current loop, being a part of the beam itself, is immersed in a magnetic field, thus a Lorentz force $\overline{F}$ arises, acting and causing (or compensating) its bending (Fig. 13.21).

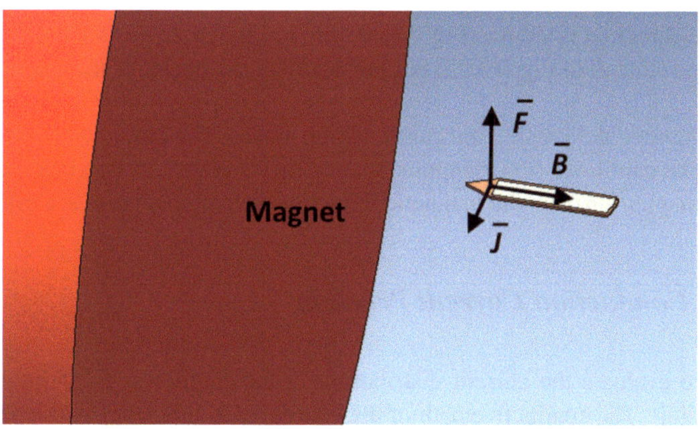

Fig. 13.21 Cantilever immersed in the magnetic field generated by the magnet

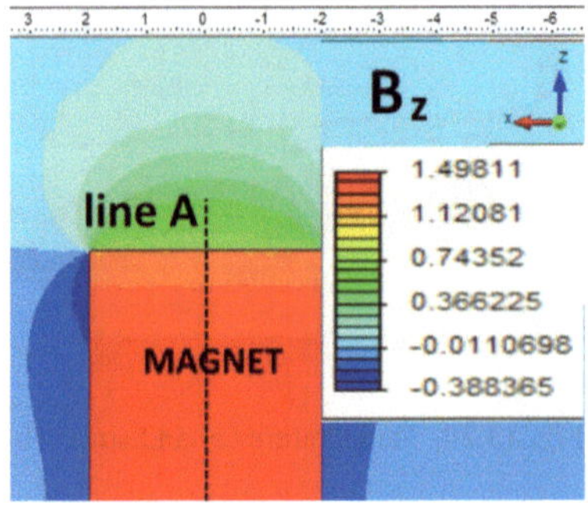

Fig. 13.22 Induction field [T] map in z-direction

The idea of the electromagnetic actuation has been described in the literature, presenting its various application [7, 22].

In this example, the impressed current is equal to 1 A and the magnetic field is generated by means of a NdFeB magnet.

The magnet is 8.36 mm long and has a radius of 4 mm. Its relative magnetic permeability is $\mu_r = 1.1$ and its coercitivity is $H_c = -1185$ kA/m.

In order to evaluate the magnetic field of the NdFeB magnet, a 3D finite element model is built and a magnetostatic problem is solved (Sect. 13.8.1), while for evaluating the current field distribution, a 2D finite element model can be used (Sect. 13.8.2). Finally, analytical formulas are given for the electro-mechanical properties of the cantilever (Sect. 13.8.3).

13.8.1 Magnetostatic Problem

A magnetostatic field problem has been solved, with Dirichlet boundary conditions applied on the boundaries of the domain.

The results are shown in Fig. 13.22, where the field map of the magnetic induction B_z along z-direction is shown (Fig. 13.22) and in Fig. 13.23 the field B_z is evaluated along a line (line A in Fig. 13.22) passing through the center of the magnet, axially located.

The magnetic field inside the cantilever tip varies between 0.46 and 0.49 T. The center of the cantilever tip (triangular shaped tip) is placed at a distance of 0.8 mm from the magnet surface: the magnetic induction field at that point is about 0.475 T.

13.8.2 Conduction Current Problem

In order to evaluate the current distribution in the cantilever, a 2D finite element model is built. The conductive path of the cantilever is simulated as copper and it is fed by a 1 mA current; a steady-state conduction field problem is solved. In Fig. 13.24 the current density distribution J_x in x-direction is shown.

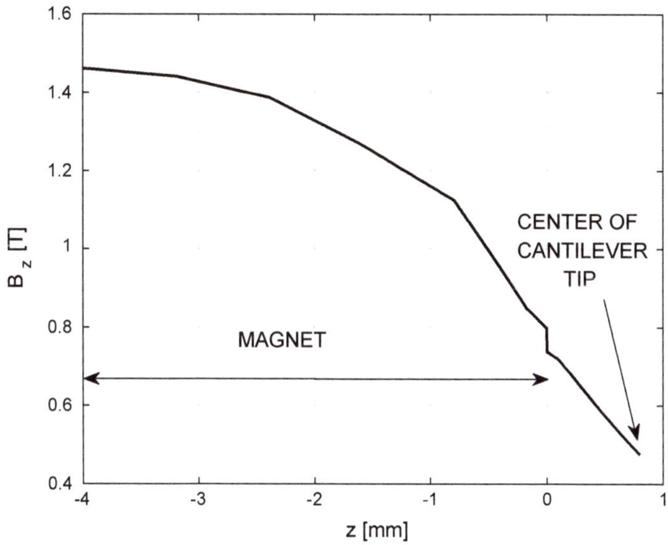

Fig. 13.23 Induction field in z-direction along line A (Fig. 13.22)

Fig. 13.24 Current density field map in x-direction. The measurement unit of current density J_x is (A m^{-2})

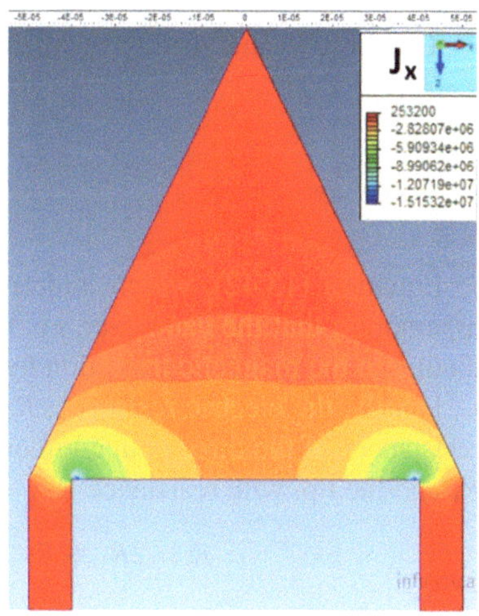

Under the hypothesis of induction field constant and equal to 0.475 T, it is possible to calculate the Lorentz force in y-direction (perpendicular to the plane x-z, where the cantilever lays) as follows:

$$F_y = B_z \int_S J_x dS = 0.475 \int_S J_x dS = 42.7\ nN \tag{13.39}$$

The integral in (13.39) can be computed numerically by finite element analysis.

13.8.3 Electro-Mechanical Properties

Given the shape g of the cantilever end, the current I, and the magnetic induction B, it is possible to analytically calculate the stiffness k of the cantilever, the resonance frequency f of the cantilever, the force F_z acting on the end region and its displacement Δz and the electric resistance R (power-loss related) of the Lorentz loop.

The stiffness k of the cantilever can be calculated as follows [22]:

$$k = \frac{Ebwt^3}{2b(L_1^3 - L_2^3) + 4wL_2^3} \quad (\mathrm{Nm}^{-1}) \tag{13.40}$$

where E is the Young's modulus, b is the cantilever width, w the arm width, t is the thickness equal to 1.5 μm, L_1 is the cantilever length and L_2 is the tip length (Fig. 13.25).

The resonance frequency f can be evaluated with the following approximated formula:

$$f \cong 0.161 \; \frac{t}{L_1^2}\sqrt{\frac{E}{\rho}} \quad (\mathrm{rad\ s}^{-1}) \tag{13.41}$$

where ρ is the mass density equal to 2330 kg m^{-3}.

The force F_z and its displacement Δz can be calculated as follows:

$$F_z = IbB \tag{13.42}$$

$$\Delta z = \frac{F_z}{k} \tag{13.43}$$

Equation (13.42), which is derived form the Lorentz's equation, is under the assumption that the cantilever, i.e. the plane in which the current flows, is perpendicular to the magnetic induction field.

Finally, the electric resistance R can be calculated as the series of three electric resistances of the three path components (two arms, with the same resistance value R_1 and the tip, with resistance R_2):

$$R = 2R_1 + R_2 \cong 2\frac{\sigma^{-1}(L_1 - L_2)}{wt} + \frac{\sigma^{-1}b}{L_2 t} \tag{13.44}$$

where σ is the electric conductivity of the boron-doped silicon (without metal layer) equal to 6.67×10^4 S m^{-1}.

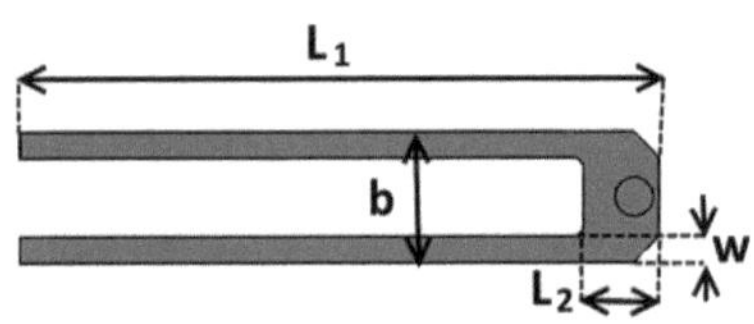

Fig. 13.25 Geometry of the cantilever

References

1. Bart SF, Mehregany M, Tavrow LS, Lang JH, Senturia SD (1992) Electric micromotor dynamics. IEEE Trans Electron Devices 39(3)
2. Chereches R, Di Barba P, Topa V, Purcar M, Wiak S (2013) Optimal shape design of electrostatic microactuators: a multiobjective formulation. Int J Appl Electromagn Mech IJAEM 43(1–2):65–76
3. Chereches RL, Di Barba P, Wiak S (2015) Non-linear inverse problems and optimal design of MEMS. COMPEL—Int J Comput Math Electr Electron Eng 34(3):608–623
4. Delinchant B, Rakotoarison HL, Ardon V, Chabedec O, Cugat O (2009) Gradient based optimization of semi-numerical models with symbolic sensitivity: application to simple ferromagnetic MEMS switch device. Int J Appl Electromagn Mech IJAEM 30:189–200
5. Di Barba P, Liu B, Mognaschi ME, Venini P, Wiak S (2017) Multiphysics field analysis and evolutionary optimization: design of an electro-thermoelastic microactuator. Int J Appl Electromagn Mech 54(3):433–448
6. Di Barba P, Dughiero F, Mognaschi ME, Savini A, Wiak S (2016) Biogeography-inspired multiobjective optimization and MEMS design. IEEE Trans Magn 52(3)
7. Di Barba P, Gotszalk T, Majstrzyk W, Mognaschi ME, Orłowska K, Wiak S, Sierakowski A (2018) Optimal design of electromagnetically actuated MEMS cantilevers. Sensors (Switzerland) 18(8)
8. Di Barba P, Mognaschi ME, Savini A, Wiak S (2016) Island biogeography as a paradigm for MEMS optimal design. Int J Appl Electromagn Mech IJAEM 51(s1):97–105
9. Di Barba P, Mognaschi ME, Venini P, Wiak S (2017) Biogeography-inspired multiobjective optimization for helping MEMS synthesis. Arch Electr Eng 66(3):607–623
10. Di Barba P, Savini A, Wiak S (1994) 2-D numerical simulation of electrostatic micromotor torque. In: Proceedings of the second international conference on computation in electromagnetics, Nottingham, pp 227–230
11. Di Barba P, Savini A, Wiak S (2008) Field models in electricity and magnetism. Springer, Berlin
12. Di Barba P, Savini A, Wiak S (2017) Higher-order multiobjective design of MEMS. Int J Appl Electromagn Mech 53(S2):S239–S247
13. Di Barba P, Wiak S (2015) Evolutionary computing and optimal design of MEMS. IEEE/ASME Trans Mechatron 20(4):1660–1667
14. Fan LS, Tai YC, Muller R (1989) IC processed electrostatic microactuator's. Sens Actuators 20:41–47
15. Fan LS, Tai YC, Muller R (1989) IC processed electrostatic synchronous microactuators. Sens Actuators 20:49–55
16. Guckel H (1998) Progress in magnetic microactuators. Microsyst Technol 5(2):59–61
17. Guckel H, Earles T, Klein J, Zook JD, Ohnstein T (1996) Electromagnetic linear actuators with inductive position sensing. Sens Actuators A 53:386–391
18. Huang QA, Lee NKS (1999) Analytical modeling and optimization for a laterally-driven polysilicon thermal actuator. Microsyst Technol 5:133–137
19. Hussein H, Tahhan A, Le Moal P, Bourbon, G, Haddab Y, Lutz P (2016) Dynamic electro-thermo-mechanical modelling of a U-shaped electro-thermal actuator. J Micromech Microeng 26(2)
20. Kolesar ES, Allen PB, Howard JT, Wilken JM, Boydston N (1999) Thermally actuated cantilever beam for achieving large in-plane mechanical deflections. Thin Solid Films 355:295–302
21. Legtenberg R, Groeneveld AW, Elwenspoek M (1996) Comb-drive actuators for large displacements. J Micromech Microeng 6:320–329
22. Majstrzyk W, Mognaschi ME, Orłowska K, Di Barba P, Sierakowski A, Dobrowolski R, Grabiec P, Gotszalk T (2018) Electromagnetic cantilever reference for the calibration of optical nanodisplacement systems. Sens Actuators A 282:149–156
23. Mehregany M, Senturia SD, Lang JH, Nagarkar P (1992) Micromotor fabrication. IEEE Trans Electron Devices 39:2060–2069

24. Paratte L, Racine GA, De Rooij NF, Bornand E (1991) Design of an integrated electrostatic stepper motor with axial field. Sens Actuators A 25–27:597–603
25. Senturia SD, Harris RM, Johnson BP, Nabors SK, Shulman MA, White JK (1992) A computer-aided design system for microelectromechanical systems (MEMCAD). J Microelectromech Syst 1:3–13
26. Tang WC, Lim MG, Howe RT (1992) Electrostatic comb drive levitation and control method. J Microelectromech Syst 1:170–178
27. Wiak S, Smołka K (2009) Numerical modelling of 3-D comb drive electrostatic accelerometers structure (method of levitation force reduction). COMPEL 28:593–602

Chapter 14
Numerical Case Studies: Inverse Problems

Maria Evelina Mognaschi

14.1 Introduction

Optimization plays a key role in the design of any device or system, and this is especially true for MEMSs. The issue is to find a design space for a device which will satisfy the performance specifications. Often, they include several design criteria which cannot all be met at the same time. This leads to the concept of multi-objective optimization, i.e., a search which attempts to satisfy several goals simultaneously. The theoretical background is based on the Pareto optimality theory [4] which was presented in Sect. 11.2.

While the basic concept of optimization, discussed in Sect. 11.1, —i.e., find the minimum or maximum value of an objective function dependent on a set of variables—is fairly obvious, in this Chapter, examples of multi-objective optimization of MEMS devices are shown.

In particular, a shape design problem is solved for all the examples presented and, depending on the complexity of the shape to improve, the number of design variables can vary, as shown in the examples in Sect. 14.2.

Moreover, solving an optimization problem for the design of a MEMS device is critical in terms of objective functions because, in order to evaluate them, electromagnetic analysis, based on two—(see devices in Sects. 14.2 and 14.4) or three-dimensional (see devices in Sects. 14.3 and 14.6) finite-element models, is needed.

This task can be computationally expensive, in particular when the field analysis implies the numerical solution to costly problems, e.g., multiphysics problems, like the one solved in Sect. 14.6. When a simplified analytical model can be used, like in the examples in Sects. 14.5 and 14.7, the problems turn out to be far less expensive.

In all the cited examples, the multi-objective problems take into account two objective functions. In the last years, attention is paid to multi-objective optimization with more than two objectives, also called many-objective optimization [10]. This happens because automated-optimal design problems are characterized by an increasing dimensionality of the objective space, as the complexity of simulation models increases. An example of many-objective optimization is shown in Sect. 14.7.

P. Di Barba and S. Wiak, *MEMS: Field Models and Optimal Design*, Lecture Notes in Electrical Engineering 573, https://doi.org/10.1007/978-3-030-21496-8_14

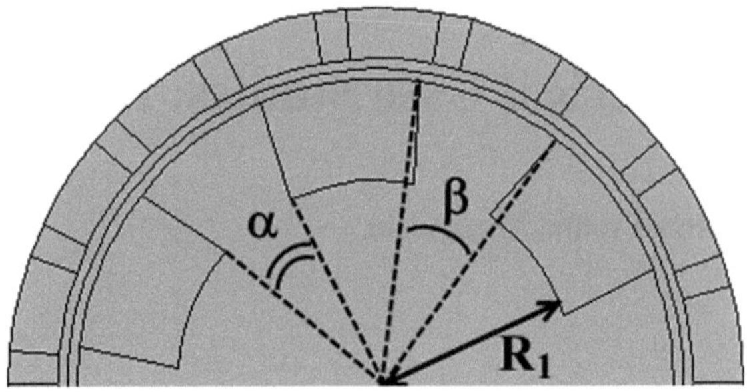

Fig. 14.1 Design variables for the optimization problem

14.2 Shape Optimization of the Electrostatic Micromotor

Because one of the main drawbacks of electrostatic micromotors is the static friction, an inverse problem which takes into account the side-pull effect has been considered. The latter is caused by the radial displacement of the rotor during its motion. Actually, the eccentric motion determines an unbalanced electric pull, which appears as a radial force acting in the direction of the shortest air-gap.

Based on Maxwell's stress tensor method, the radial force $F_r(\varphi)$ acting on the rotor can be evaluated as a function of its angular position φ, in addition to the driving torque $T_d(\varphi)$.

In the inverse problem here considered, the shape of the electrostatic micromotor described in Sect. 13.3.2 is optimized; to simulate the radial displacement, a clearance between rotor and shaft equal to 1 μm was considered; the direction of displacement was assumed to be fixed and independent of φ (static eccentricity).

14.2.1 Three Design Variables Optimization

The shape design of the rotor has been considered, taking radius R_1 and angles (α, β) as the design variables, as shown in Fig. 14.1 [9].

Significantly, two objective functions can be defined in terms of design vector $g = (R_1, \alpha, \beta)$ namely:

- the highest value of driving torque on a pole pitch $f_1(g)$ at no load under single-phase supply;
- the value of radial force on the rotor $f_2(g)$ in the direction of the shortest air-gap.

The inverse problem reads: given stator supply and rotor misalignment, find the family of rotor geometries g such that $f_1(g)$ is maximum and $f_2(g)$ is minimum according to the Pareto definition of non-dominated solution.

The BiMo method presented in Sect. 11.6 is used. It has been applied with mutation probability set to 0.04; the run was stopped after 5 generations, each based on 14

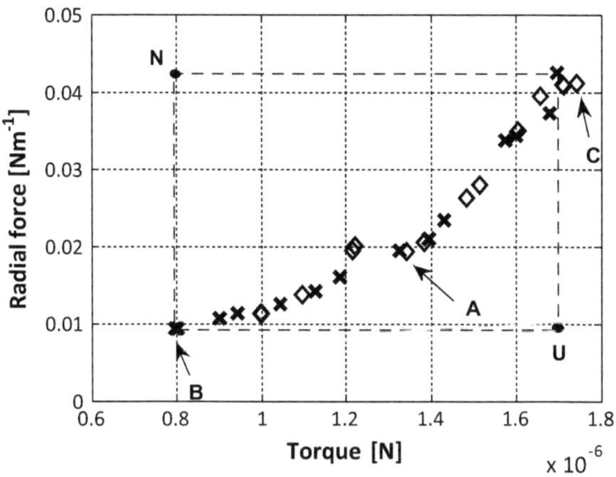

Fig. 14.2 Objective space and non-dominated solutions: cross—NSGA-II results, diamond—BiMO multi-objective results

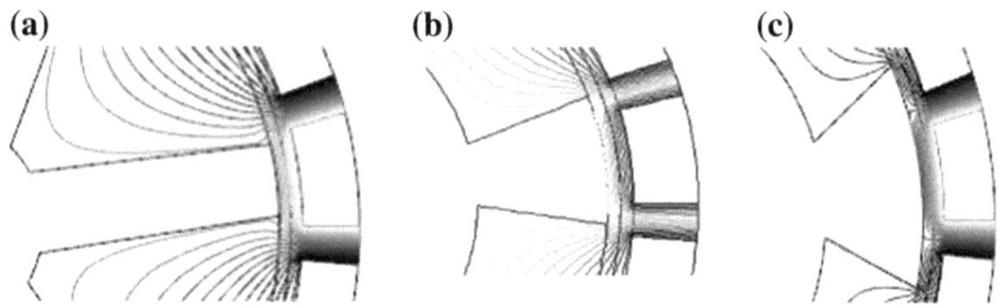

Fig. 14.3 Geometry of solution **a** (left), **b** (middle) and **c** (right) of Fig. 14.2

Table 14.1 Results of A, B and C solutions in Fig. 14.2

	R_1 (μm)	α (°)	β (°)	Torque (μN)	Radial force (mN m^{-1})
Sol. A	18.6	83.4	10.0	0.80	9.30
Sol. B	39.8	11.6	23.1	1.34	19.4
Sol. C	41.2	11.6	40.8	1.74	41.1

habitats. For the sake of a comparison, the front was also approximated by means of NSGA-II. This last was stopped after 5 generations, each based on 14 individuals.

The approximated Pareto fronts are shown in Fig. 14.2, where also the Utopia U and Nadir N points are highlighted.

In Fig. 14.3 geometry and potential lines of non-dominated solutions A, B and C of Fig. 14.2 are shown.

These solutions are characterized by design variable values and objective function values shown in Table 14.1.

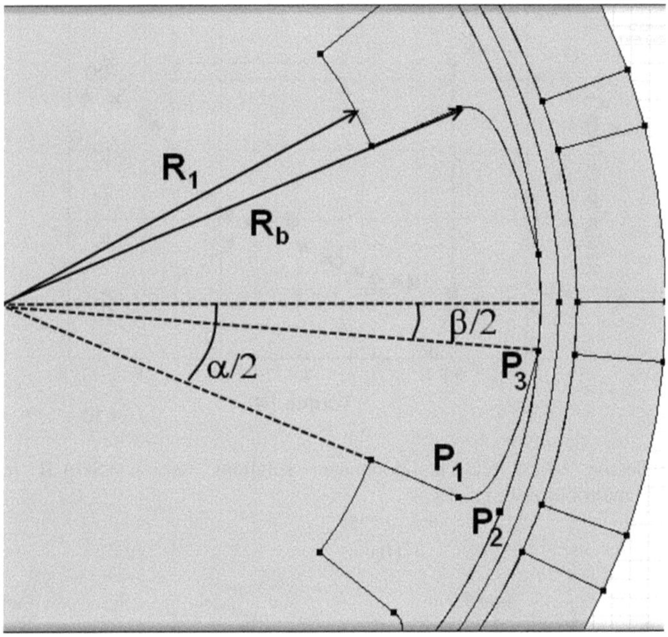

Fig. 14.4 Control points and design variables of the optimization problem

Table 14.2 Bounds of the design variables

	x_1 (μm)	x_2 (μm)	x_3 (°)	x_4 (°)	x_5
min	15	15	10	10	0.01
max	55	60	55	55	10

14.2.2 Five Design Variables Optimization

In this optimization, a device with the shape of tooth corner (Fig. 14.4) rounded by means of a second-order Bézier curve is considered [11]. The control points of the Bézier curve exhibit polar coordinates $P_1(R_b, -\alpha/2)$, $P_2(R_2, -\alpha/2)$ and $P_3(R_2, -\beta/2)$. A weight $k > 0$ makes the tooth corner filled in.

In the inverse problem, variables $(R_1, R_b, \alpha, \beta, k)$ have been selected as the entries of the five dimensional design vector g. The variation range for the design variables are shown in Table 14.2.

The outer radius of the rotor R_2 is constant and equal to 60 μm.

Accordingly, two objective functions have been defined, namely:

- the highest value of driving torque $f_1(g)$ under single-phase supply at no load;
- the value of radial force on the rotor $f_2(g)$ in the direction of the shortest air-gap.

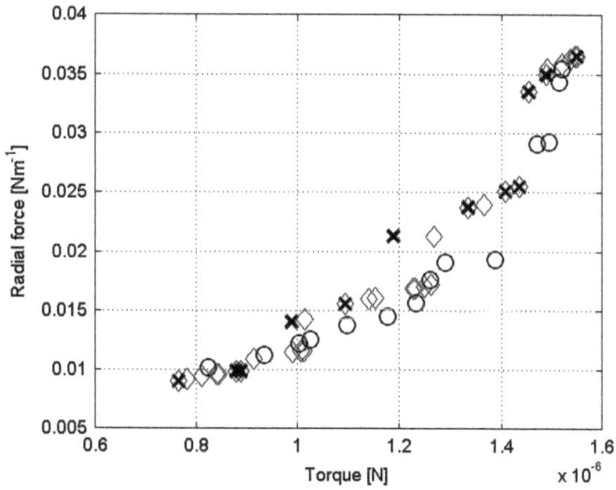

Fig. 14.5 Objective space and non-dominated solutions: circle—NSGA-II, cross—non-linear BBO, diamond—non-linear BBO with storage of habitats

The optimal design problem reads: given stator supply and rotor misalignment, find the family of rotor geometries g such that $f_1(g)$ is maximum and $f_2(g)$ is minimum according to the Pareto definition of non-dominated solution.

The BiMO algorithm (see Sect. 11.6) is applied with a non-linear model of the immigration curve [11].

The emigration rate μ is calculated as $\mu = 1 - \lambda$.

In Fig. 14.5 a comparison between non-linear BBO and NSGA-II [3] methods (converged after $n_g = 10$ generations, based on $N_p = 14$ solutions each) is shown. Moreover, at each generation of the BBO optimization, the habitats were stored: at the end, the Pareto front numerically resulting from all the habitats stored by BBO was considered. Storing the solutions found at each iteration is costless and allows to consider good solutions which have been lost in the procedure. This way, the social behavior of BBO is fully exploited.

14.3 Shape Optimization of a Comb-Drive Accelerometer

The comb-drive accelerometer described in Sect. 13.4 is here considered for an optimal shape design [1, 15].

A comb-drive actuator needs to be as coplanar as possible with respect to its sets of movable and fixed electrodes. To this end, various solutions have already been considered [16]. In the case study, a grounded substrate is laid under the set of electrodes as an attempt to cancel the vertical force. This way, however, the electric field distribution is no longer symmetric and the movable electrodes tend to levitate when the comb drive is energized. This vertical perturbation must be kept as low as possible while simultaneously the drive force should be increased. Therefore, the ultimate goal of the optimal shape design problem is to find the family of geometries that maximize the x-directed drive force between movable and fixed electrodes, and

Table 14.3 Details of two non-dominated solutions

w_m (μm)	w_f (μm)	h_m (μm)	h_f (μm)	F_x (N)	Slope of F_z versus z (N/μm)
6	6	6.2	6.1	3.88×10^{-10}	2.79×10^{-10}
7.7	7.8	7.7	7.8	5.86×10^{-10}	4.03×10^{-10}

simultaneously minimize the z-directed levitation force: a bi-objective optimization problem is so originated. To this end, the four-dimensional vector $g = (w_m, w_f, h_m, h_f)$ of design variables has been defined; they are discrete valued (step 0.1 μm) and can range from 2 to 8 μm. Moreover, a two-dimensional vector $F = (f_1, f_2)$ of objective functions has been defined, such that:

- drive $f_1(g) = F_x(x, g)$ for $z = 0$ and $0 \leq x \leq 14$ μm, to be maximized with respect to vector g;
- levitation $f_2(g) = F_z(z, g)$ for $x = -13$ μm and $0 \leq z \leq 1.8$ μm, to be minimized with respect to vector g.

In practice, the average value of the F_x versus x curve is maximized, and simultaneously the slope of the F_z versus z-displacement curve is minimized against vector g.

The automated optimal design is based on a surrogate model: connecting the MATLAB surrogate modeling (SUMO) toolbox with COMSOL Multiphysics finite element analysis tool, the surrogate model is derived. Then, the NSGA-II algorithm [3], a well known optimization algorithm, is used for solving this optimization problem, based on the surrogate model.

A set of sixty-one non-dominated solutions approximate the Pareto front [2]; for the sake of an example, the detail of two non-dominated solutions is reported in Table 14.3.

14.4 Shape Optimization of a Micro-Mirror Actuator

The magnetic MEMS shown in Sect. 13.5 is here optimized. In the following Sections a bi-objective formulation [15] as well as a tri-objective formulation [14] of the optimization problem are proposed and solved.

14.4.1 Bi-Objective Optimization Results

The design problem is set up as follows: having prescribed lower thresholds for holding and actuation torque, given the plate angle $\varphi = 10°$, find the geometry of

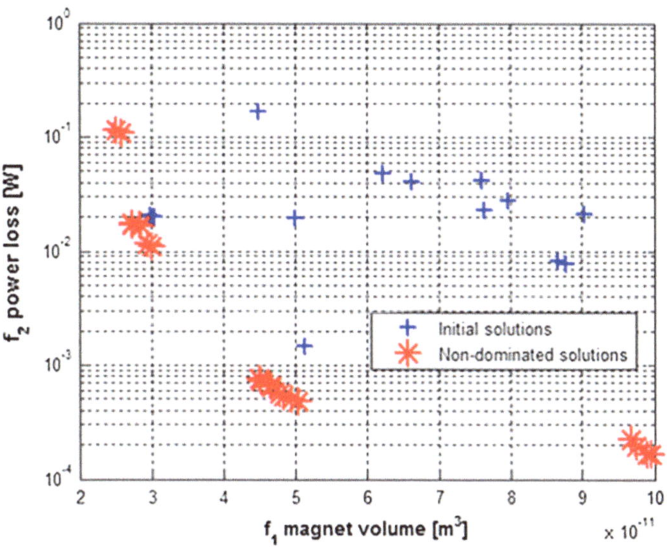

Fig. 14.6 Optimization results. Constraints: holding torque >1 nN m, actuation torque >0.25 nN m

magnet and conductors, as well as the amplitude of the current pulse, such that the power loss in the conductors and the magnet volume are both minimized.

From the viewpoint of the computational cost, it can be remarked that the evaluation of magnet volume $x_1 x_2$ and power loss $\omega x_7^2 (\sigma x_4 x_5)^{-1}$ with $\sigma = 6 \times 10^7$ S m^{-1} is straightforward, while the evaluation of both holding and actuation torque at $\varphi = 10°$ is field dependent. In order to deal with feasible design configurations, suitable geometric constraints have been considered; moreover, the upper bound $x_7 (x_4 x_5)^{-1} < 5 \times 10^9$ A m^{-2} have been fulfilled for the pulsed current density.

Moreover, the lower threshold for the holding torque is set to 1 nN m, while for the actuation torque it is set to 0.25 nN m.

The NSGA-II method [3] is run with 20 individuals and 40 generations.

An approximation of the Pareto front in the power-loss versus magnet-volume space is shown in Fig. 14.6.

14.4.2 Tri-Objective Optimization Results

While in Sect. 14.4.1 an optimization problem with two design criteria was investigated, here a 3D objective space is considered.

The optimal design problem reads: having prescribed lower thresholds for holding torque (10^{-8} Nm) and actuation torque (10^{-9} Nm), given mirror position $\varphi = 10°$, find the geometry of magnet and conductors, as well as the current pulse amplitude, such that: power loss in conductors (f_1), weighted material cost (f_2), and excitation

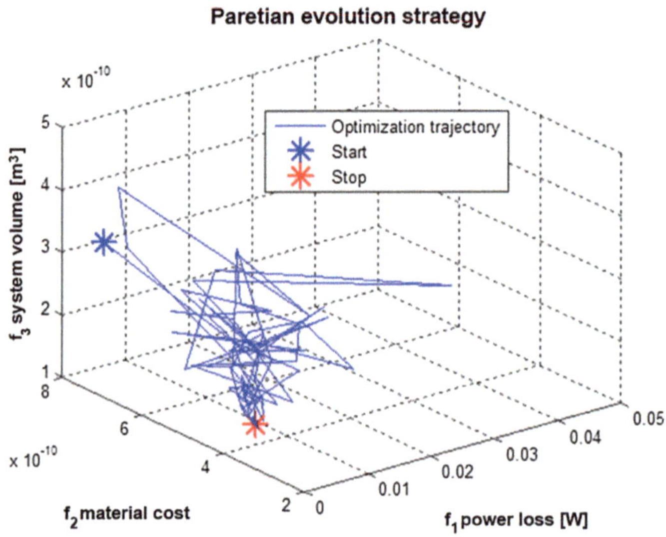

Fig. 14.7 Paretian evolution strategy (P-EStra): optimization trajectory in the objective space

Table 14.4 P-EStra: Objective improvement $f(\text{stop})/f(\text{start})$ and design variable movement $x(\text{stop})/x(\text{start})$ for the optimal point in Fig. 14.7

f_1	f_2	f_3	x_1	x_2	x_3	x_4	x_5	x_6	x_7
0.218	0.444	0.507	1.098	0.841	0.159	1.066	0.902	0.388	0.298

system volume (i.e. conductors and magnet, f_3) are simultaneously minimized in the Pareto sense. In a way, objectives f_1 and f_3 are the main conflict pair because in the MEMS domain miniaturisation is limited by the thermal behaviour of components.

The P-EStra algorithm described in Sect. 11.4.1 is used for solving this problem.

The results are shown in Fig. 14.7 and Table 14.4: it can be remarked that objective functions and design variables change substantially moving from the initial solution to the final one.

14.5 Shape Optimization of a Two-Arms Electro-Thermo-Elastic Microactuator

The geometry of the two-arms microactuator described in Sect. 13.6 is here optimized: the design variables are L_1, L_2, L_g, h_h and d [12].

The objective functions are the maximum temperature T_m to be minimized, and the displacement U, which is the deflection in the free end of the actuator, to be maximized.

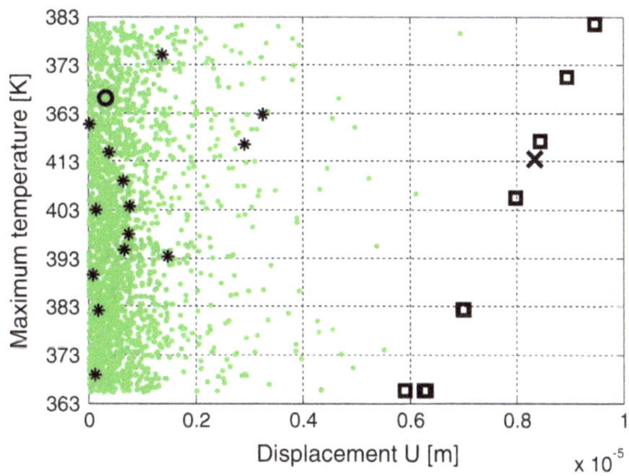

Fig. 14.8 Optimisation results: star—starting points BiMO, square—arrival points BiMO, circle—starting point goal attainment, cross—arrival point goal attainment, dot—random sampling

Two different algorithms have been applied: goal attainment method [4] and the biogeography-based method (see Sect. 11.6).

Goal attainment is a well-known deterministic method, which allows to solve a multi-objective problem by reformulating it in terms of a minimax problem, subject to suitable weights expressing the user preferences. The minimax problem is subsequently solved by means of a deterministic algorithm, i.e. the sequential quadratic programming [18].

The BiMO algorithm has been applied with the following stopping criterion: the computation stops when mean $m_{f,\, mean}$ and maximum $m_{f,\, max}$ values of m_f over the last five iterations fulfill the following condition: $m_{f,\, mean} < 1\%$ and $m_{f,\, max} < 2\%$.

The starting points are randomly generated.

Goal attainment method has been used with the default settings for the routine *fgoalattain* of MatLab [18]. The point $[-10^{-6}, 369]$ is taken as the goal vector, while the weights are the absolute value of the two components of the goal vector.

The optimisation results are shown in Fig. 14.8, where also a random sampling of the search space is shown (dots).

The starting point for the goal attainment algorithm is the star while the arrival point is the cross. Goal attainment is able to find a solution belonging to the Pareto front. The run lasted 75 iterations (533 objective function calls).

BiMO algorithm is able to find a good approximation of the whole front. In fact, 13 points out of 14 converged to the approximated front. The run lasted 806 iterations (11284 objective function calls).

In order to approximate the whole front with the goal attainment algorithm, it would be necessary to run many times the algorithm, each time varying its weights.

It is possible to say that both methods find Pareto-optimal solutions with hundreds of objective function calls per individual.

Moreover, the computational cost is comparable, if one considers that BiMO takes more time but finds 14 solutions, while goal attainment finds out just one solution. The computation time could be substantially high when considering numerical models for solving the forward problem.

Also the displacements found for both devices exhibit the same order of magnitude.

14.6 Shape Optimization of a Three-Arms Electro-Thermo-Elastic Microactuator

A shape optimization of the device analyzed in Sect. 13.7 is here optimized [5]. The problem reads as follows: acting on the following design variables

L length of the actuator,
h_h thickness of the actuator,
d_w width of the cold arm,
d width of the hot arms,

two objective functions are defined:

$$f_1 = u|_{V=V\,min} + u|_{V=V\,max} \tag{14.1}$$

$$f_2 = T_{max}|_{V=V\,min} + T_{max}|_{V=V\,max} \tag{14.2}$$

where f_1 is to be maximized and f_2 to be minimized, with V_{min}= 1 V, V_{max}= 5 V, subject to the constraints:

$$T_{max} < 1500 \text{ K} \tag{14.3}$$

$$\|S\| < 1.44 \text{ GPa}. \tag{14.4}$$

The meaning of the (14.4) is that the norm of the stress at the most solicited point of the actuator should not exceed the silicon breaking stress.

The design variable ranges are L [56–300] μm, h_h [2–5] μm, d_w [7–30] μm, d [1–7] μm, respectively.

The BiMO algorithm (see Sect. 11.6) is applied for solving this optimization problem. Considering the good performance of the elitism technique applied in the study of the method [11], half of the Pareto front is considered for elitism. In fact, using half of the Pareto front for elitism is suitable when a time-consuming forward problem (like (13.27)–(13.38) is in 3D) has to be solved at each iteration.

An ecosystem composed of $n_i = 14$ islands is considered.

In order to rule the convergence of the proposed method, the following index is considered:

$$m_f(k) = \frac{|d_f(k) - d_f(k-1)|}{d_f(k-1)} \tag{14.5}$$

with

$$d_f = \frac{1}{N_p}\sqrt{\sum_{i=1}^{N_p}\sum_{j=1}^{N_f}\left(\frac{f_{i,j} - u_j}{L_j}\right)^2}, \tag{14.6}$$

where N_p is the number of habitats, N_f is the number of objective functions, L_j is the distance between Utopia and Nadir points while $f_{i,j}$ and u_j are the current values of the objective functions and the utopia point, respectively. Put simply, index d_f is related to the mean distance of the current habitats to the Utopia point while index m_f is calculated at the k-th iteration and gives information about how much the mean distance d_f to Utopia changed at the k-th iteration with respect to the previous one.

The stopping criterion is based on a twofold condition as follows: the optimisation stops when the mean value of m_f considering the last 5 iterations, $m_{f,\,mean}$, is smaller than a prescribed threshold, e.g. 0.01 and the maximum value of m_f considering the last 5 iterations, $m_{f,\,max}$, is smaller than a prescribed threshold, e.g. 0.02.

The optimisation lasted n_g= 35 iterations. At each iteration the Pareto front is calculated and archived, considering also the history of the optimisation results: the Pareto front so found is shown in Fig. 14.9. For each method the overall cost is proportional to $n_i \times n_g$ calls to the FEA. Each call to FEA lasts, depending on the mesh discretizing the geometry, from 5 through 20 min on a computer equipped with a processor Intel i7, 3.6 GHz and 16 GB of RAM.

In order to make a comparison between the results found for the two-arms microactuator (Sect. 14.5) and the three-arms microactuator, it can be stated that the maximum value of temperatures for the two Pareto fronts are comparable. It can be noted

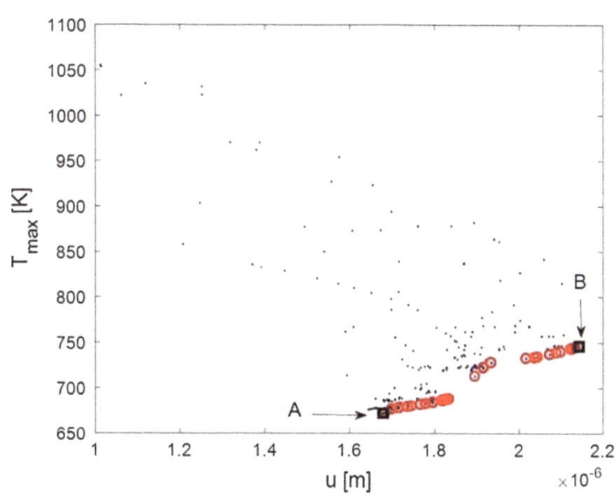

Fig. 14.9 Objective space (dot) and non-dominated solutions (circle); A and B points (square) are the ends of the approximated Pareto front

from (14.2) that the objective function f_2 is the sum of two temperatures; hence, the temperature values in Fig. 14.8 have to be approximately doubled in order to be compared with results in Fig. 14.9. The temperature maps of solutions A and B (Fig. 14.9) are shown in Fig. 14.10.

Two solutions, located at the end points of the Pareto front, are considered. Results in terms of design variable, objective function and constraint values are shown in Table 14.5.

Taking into account the results in Table 14.5, it can be stated that, from the designer point of view, a long actuator should be preferred. Moreover, considering the bounds of the height h_h, it can be observed that the value of h_h is the smallest one for device A and the highest one for device B. A further role is played by the ratio d_w to d, which should be high. Therefore, according to extra preferences of the device designer, a final choice can be made a posteriori among Pareto-optimal solutions.

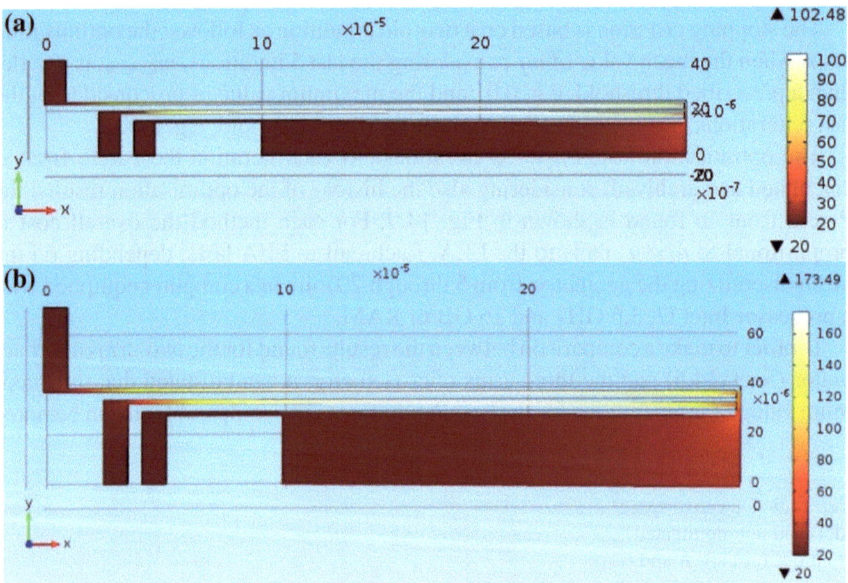

Fig. 14.10 Temperature [°C] maps of solutions A (**a**) and B (**b**)

Table 14.5 Results of the two end-points of the Pareto front in Fig. 14.9

Design variables (μm)	A	B	Objective function/constraint	A	B
L	293.75	286.9	f_1 (μm)	1.68	2.14
h_h	2.03	4.97	f_2 (K)	672.2	746.3
d_w	15.76	29.13	T_{max} (K)	375.6	446.6
d	1.00	1.72	S (GPa)	0.19	0.44

Table 14.6 Variation range for the design variables (units in μm)

	w	L_1	L_2	b
Lower bound	20	100	50	100
Upper bound	–	600	100	150

In order to compare this method with a well-known classical method, the goal attainment algorithm (the same used for the two-arms microactuator, Sect. 14.6) has been resorted to. However, after many numerical experiments, it was not able to find any non-dominated solution on the Pareto front. This is probably due to the non-smoothness of objective functions (14.1–14.2), which makes the problem stiff for a deterministic method like goal attainment.

14.7 Shape Optimization of the Electromagnetically Actuated Cantilever

A shape optimization of the device shown in Sect. 13.8 is here optimized [17].

If the shape of the cantilever end is defined by means of a n-dimensional vector $g = (g_1, \dots, g_k, \dots, g_n)$ of geometric variables (e.g. for a polygonally-shaped end region, the coordinates of the relevant vertices), the inverse (or design) problem reads: given current I and magnetic induction B, find the shape $g = (g_1, \dots, g_k, \dots, g_n)$ of the cantilever end region such that:

the stiffness $k(g)$ of the cantilever is minimized;
the resonance frequency $f(g)$ is maximized;
the displacement $\Delta z(g)$ of the end region is maximized;
the electric resistance $R(g)$ of the Lorentz loop is minimized.

A multi-objective optimisation problem characterized by four objective functions $[k(g), f(g), \Delta z(g), R(g)]$ is originated. When many objective functions (say more than two) are considered, it is very common to find solutions of the optimization problem which are equivalent in the Pareto sense to the starting point. However, these solutions are nevertheless interesting because they improve at least one objective function.

The shape of the cantilever is defined by four design variables, as shown in Fig. 13.24: w, arm width, L_1, cantilever length, L_2, tip length, b, cantilever width [7].

The variation range for each design variable is shown in Table 14.6.

In order to guarantee a geometrical consistency, the following constraint (units in μm) is set:

$$b \geq 2w + 10 \tag{14.7}$$

Subsequent optimization runs are considered, with one (Opt1), two (Opt2) or more (Opt3) objective functions at a time:

Opt1—each objective function i.e. k, f, Δz and R, is optimized in four different single-objective optimizations (Opt1k, Opt1f, Opt1z, Opt1R);
Opt2—the following pairs of functions are considered: k and R are optimized (Opt2kR), f and z are optimized (Opt2fz), f and R are optimized (Opt2fR), z and R are optimized (Opt2zR);
Opt3—k, z and R are simultaneously optimized (Opt3).

In order to solve these optimization problems, the evolutionary algorithm of lowest order described in Sect. 11.4 is applied [4]. This algorithm is able to solve single-objective problems (in this case it is called "EStra method" (Di Barba, Savini and Wiak, [8, 13], multi-objective problems ("MOESTRA method" [6], see Sect. 11.4.1) and many-objective problems ("P-EStra method", see Sect. 11.4.1).

14.7.1 Single-Objective Optimization Results

The results of the single-objective optimizations are shown in Table 14.7. In each table the values of the design variables and of the functions (k, f, Δz, R) are shown. In particular, the minimized objective function is highlighted in bold.

From Table 14.7—Opt1k, it can be noted that the non-controlled objective functions f decreases, which is unwanted, Δz increases (desiderable) and R increases (unwanted).

From Table 14.7—Opt1f, it can be noted that the non-controlled objective functions k increases, which is unwanted, Δz decreases (unwanted) and R decreases (desiderable).

From Table 14.7—Opt1z, it can be noted that the non-controlled objective functions k decreases (desiderable), f decreases (unwanted) and R increases (unwanted).

From Table 14.7—Opt1R, it can be noted that the non-controlled objective functions k increases (unwanted), f increases (desiderable) and Δz decreases (unwanted).

Table 14.7 Single-objective optimization Opt1k results

	w (μm)	L_1 (μm)	L_2 (μm)	b (μm)	k (Nm^{-1})	f (kHz)	Δz (nm)	R (Ω)
Initial	20	500	50	100	4.32×10^{-2}	8	925.37	470
Opt1k	21.05	568.92	53	114.39	**3.09×10^{-2}**	6.18	1481.94	511.84
Opt1f	24.55	210.26	64.57	122.56	0.726	**45.26**	67.56	137.67
Opt1z	22	567.57	63.93	127.90	3.25×10^{-2}	6.21	**1573.71**	478
Opt1R	47.59	210.51	80.73	119.47	1.39	45.16	34.29	**69.34**

14.7.2 Bi-Objective Optimization Results

The results of the bi-objective optimizations are shown in Table 14.8. In each table the values of the design variables and of the functions (k, f, Δz, R) are shown. In particular, the minimized objective functions are highlighted in bold.

From Table 14.8—Opt2kR, it can be noted that the non-controlled objective function f decreases, which is unwanted, but Δz increases (desiderable).

From Table 14.8—Opt2fz, it can be noted that the non-controlled objective function k increases, which is unwanted, but R decreases (desiderable).

From Table 14.8—Opt2fR, it can be noted that the non-controlled objective function k increases (unwanted) and Δz decreases (unwanted).

From Table 14.8—Opt2zR, it can be noted that the non-controlled objective function k decreases (desirable) and f decreases (unwanted).

14.7.3 Tri-Objective Optimization Results

The results of the tri-objective optimization are shown in Table 14.9. The values of the design variables and of the functions (k, f, Δz, R) are shown. In particular, the minimized objective functions are highlighted in bold.

Table 14.8 Bi-objective optimization results

	w (μm)	L_1 (μm)	L_2 (μm)	b (μm)	k (Nm^{-1})	f (kHz)	Δz (nm)	R (Ω)
Initial	20	500	50	100	4.32×10^{-2}	8	925.37	470
Opt1 k	24.05	557.13	62.33	110	**3.76×10^{-2}**	6.45	1170.59	**429.13**
Opt1f	21	490.14	55.40	136.12	4.82×10^{-2}	**8.33**	**1129.24**	438.48
Opt1z	60.92	210.04	86.10	135.02	1.79	**45.36**	30.29	**56.37**
Opt1R	27.17	568.93	57.71	110.96	3.99×10^{-2}	6.18	**1113.33**	**395.46**

Table 14.9 Tri-objective optimization Opt3 results

	w (μm)	L_1 (μm)	L_2 (μm)	b (μm)	k (Nm^{-1})	f (kHz)	Δz (nm)	R (Ω)
Initial	20	500	50	100	**4.32×10^{-2}**	8	**925.37**	**470**
Final	23.97	562.15	68.13	111.78	**3.65×10^{-2}**	6.33	**1226.19**	**428.66**

From Table 14.9, it can be noted that the non-controlled objective function f decreases, which is unwanted.

Provided that all the optimization runs start from the same initial point, the following remarks can be put forward.

The solutions obtained in Opt1k, Opt1f, Opt1z and Opt1R differ in both design vector (w, L_1, L_2, b) and objective vector (k, f, Δz, R). This proves that a single solution simultaneously satisfying all the design criteria does not exist. From the optimisation theory viewpoint, it is a design conflict problem that can be studied via Pareto optimality.

Solutions obtained in Opt2 and Opt3 can be considered Pareto-equivalent to the initial one, because three objectives improve, while one objective deteriorates (e.g. solution of Opt3 where k, Δz and R improve, while f deteriorates). From the application viewpoint, provided the amount of deterioration in one objective is acceptable, Pareto-equivalent solutions may represent a good alternative to the initial solution.

References

1. Chereches R, Di Barba P, Topa V, Purcar M, Wiak S (2013) Optimal shape design of electrostatic microactuators: a multiobjective formulation. Int J Appl Electromagnetics Mech IJAEM 43(1–2):65–76
2. Chereches RL, Di Barba P, Wiak S (2015) Non-linear inverse problems and optimal design of MEMS. COMPEL—Int J Comput Math Electr Electr Eng 34(3):608–623
3. Deb K, Pratap A, Agarwal S, Meyarivan T (2002) A fast and elitist multi-objective genetic algorithm: NSGA-II. IEEE Trans Evol Comput 6(2):182–197
4. Di Barba P (2010) Multiobjective shape design in electricity and magnetism. Springer
5. Di Barba P, Liu B, Mognaschi ME, Venini P, Wiak S (2017) Multiphysics field analysis and evolutionary optimization: design of an electro-thermoelastic microactuator. Int J Appl Electromagnet Mech 54(3):433–448
6. Di Barba P, Mognaschi ME (2009) Industrial design with multiple criteria: shape optimization of a permanent-magnet generator. IEEE Trans Magnetics 45(3):1482–1485
7. Di Barba P, Gotszalk T, Majstrzyk W, Mognaschi ME, Orłowska K, Wiak S, Sierakowski A (2018) Optimal design of electromagnetically actuated MEMS cantilevers. Sensors 18(8):2523
8. Di Barba P, Mognaschi ME, Lowther DA, Sykulski JK (2018) A benchmark TEAM problem for multi-objective pareto optimization of electromagnetic devices. IEEE Trans Magnetics 54(3):1
9. Di Barba P, Dughiero F, Mognaschi ME, Savini A, Wiak S (2016) Biogeography-inspired multiobjective optimization and MEMS design. IEEE Trans Magnetics 52(3):1
10. Di Barba P, Mognaschi ME, Rezaei N, Lowther DA, Rahman T (2019) Many-objective shape optimisation of IPM motors for electric vehicle traction. Int J Appl Electromagnetics Mech IJAEM
11. Di Barba P, Mognaschi ME, Savini A, Wiak S (2016) Island biogeography as a paradigm for MEMS optimal design. Int J Appl Electr Mech IJAEM 51(s1):97–105
12. Di Barba P, Mognaschi ME, Venini P, Wiak S (2017) Biogeography-inspired multiobjective optimization for helping MEMS synthesis. Arch Electr Eng 66(3):607–623
13. Di Barba P, Savini A, Wiak S (2008) Field models in electricity and magnetism. Springer, Berlin, Germany
14. Di Barba P, Savini A, Wiak S (2017) Higher-order multiobjective design of MEMS. Int J Appl Electromagnet Mech 53(S2):S239–S247
15. Di Barba P, Wiak S (2015) Evolutionary computing and optimal design of MEMS. IEEE/ASME Trans Mech 20(4):1660–1667

16. Legtenberg R, Groeneveld AW, Elwenspoek M (1999) Comb-drive actuators for large displacements. J Micromech Microeng 6:320–329
17. Majstrzyk W, Mognaschi ME, Orłowska K, Di Barba P, Sierakowski A, Dobrowolski R, Grabiec P, Gotszalk T (2018) Electromagnetic cantilever reference for the calibration of optical nanodisplacement systems. Sens Actuators, A 282:149–156
18. Mathworks (2019) MatLab documentation [Online]. Available at https://it.mathworks.com/help/matlab/. Accessed on 6 March 2019

Appendix

Elementary Vector Analysis

In a three-dimensional domain, using rectangular coordinates, special vectors are the space vectors $\bar{r}_P = (x_P, y_P, z_P)$ and $\bar{r}_Q = (x_Q, y_Q, z_Q)$, defining field point P and source point Q, respectively (see Fig. A.1).

The Euclidean distance between P and Q is

$$r = |\bar{r}| = \left[(x_P - x_Q)^2 + (y_P - y_Q)^2 + (z_P - z_Q)^2\right]^{\frac{1}{2}} = \left|\bar{r}_P - \bar{r}_Q\right| \tag{A.1}$$

A special function is $\frac{1}{r}, \quad r \neq 0$; it results:

$$\overline{\nabla}_P r = \frac{\bar{r}}{r}, \overline{\nabla}_Q r = -\frac{\bar{r}}{r} \tag{A.2}$$

$$\overline{\nabla}_P \frac{1}{r} = -\frac{\bar{r}}{r^3} \quad , \quad \overline{\nabla}_Q \frac{1}{r} = \frac{\bar{r}}{r^3} \tag{A.3}$$

$$\overline{\nabla}_P \cdot \bar{r} = \overline{\nabla}_Q \cdot \bar{r} = -3 \tag{A.4}$$

$$\nabla_P^2 \frac{1}{r} = -4\pi\delta(r) \text{ where } \delta \text{ denotes the Dirac's distribution} \tag{A.5}$$

$$\overline{\nabla} \times \left(-\frac{\bar{r}}{r^3}\right) = \overline{\nabla} \times \overline{\nabla} \frac{1}{r} = 0 \tag{A.6}$$

$$\int_\Omega \frac{\delta(r_Q)}{r} d\Omega = \frac{1}{r} \tag{A.7}$$

P. Di Barba and S. Wiak, *MEMS: Field Models and Optimal Design*, Lecture Notes in Electrical Engineering 573, https://doi.org/10.1007/978-3-030-21496-8

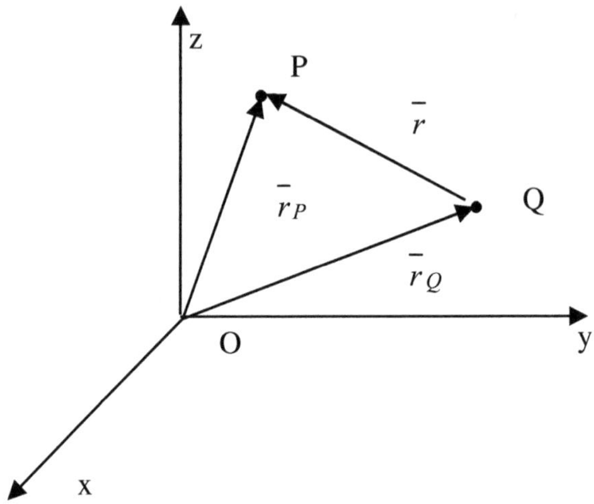

Fig. A.1 Field point P and source point Q

The following formulae hold:

$$\overline{\nabla} \cdot \left(\overline{\nabla} \times \overline{V}\right) = 0 \tag{A.8}$$

$$\overline{\nabla} \times \overline{\nabla}\phi = 0 \tag{A.9}$$

Gauss's or divergence theorem

$$\int_{\Omega} \overline{\nabla} \cdot \overline{B}\, d\Omega = \int_{\Gamma} \overline{B} \cdot \bar{n} d\Gamma \tag{A.10}$$

with $\Gamma = \partial\Omega$ closed surface enclosing Ω and $\bar{n}$ outward normal versor.

Stokes's or circulation theorem

$$\oint_{\ell} \overline{A} \cdot \bar{t}\, d\ell = \int_{\Gamma} \left(\overline{\nabla} \times \overline{A}\right) \cdot \bar{n}\, d\Gamma \tag{A.11}$$

with $\ell = \partial\Gamma$ closed line representing the border of Γ and $\bar{t}$ anticlockwise tangential versor.

Vector identities

$$\overline{\nabla} \times \left(\overline{\nabla} \times \overline{A}\right) = \overline{\nabla}\left(\overline{\nabla} \cdot \overline{A}\right) - \overline{\nabla}^2 \overline{A} \tag{A.12}$$

$$\overline{\nabla}\cdot\left(\overline{V}_1\times\overline{V}_2\right)=\overline{V}_2\cdot\left(\nabla\times\overline{V}_1\right)-\overline{V}_1\cdot\left(\nabla\times\overline{V}_2\right) \tag{A.13}$$

$$\overline{\nabla}\cdot\left(\varphi\overline{V}\right)=\overline{V}\cdot\overline{\nabla}\varphi+\varphi\overline{\nabla}\cdot\overline{V} \tag{A.14}$$

$$\overline{\nabla}\times\left(\overline{V}_1\times\overline{V}_2\right)=\overline{V}_1\overline{\nabla}\cdot\overline{V}_2-\overline{V}_2\overline{\nabla}\cdot\overline{V}_1+\left(\overline{V}_2\cdot\overline{\nabla}\right)\overline{V}_1-\left(\overline{V}_1\cdot\overline{\nabla}\right)\overline{V}_2 \tag{A.15}$$

$$\overline{\nabla}\times\varphi\overline{V}=\varphi\overline{\nabla}\times\overline{V}+\overline{\nabla}\varphi\times\overline{V} \tag{A.16}$$

In cylindrical coordinates

$$\overline{\nabla}\cdot\overline{V}=r^{-1}\frac{\partial(rV_r)}{\partial r}+r^{-1}\frac{\partial V_\vartheta}{\partial\vartheta}+\frac{\partial V_z}{\partial z} \tag{A.17}$$

$$\nabla^2U=\frac{\partial^2U}{\partial r^2}+r^{-1}\frac{\partial U}{\partial r}+r^{-2}\frac{\partial^2U}{\partial\vartheta^2}+\frac{\partial^2U}{\partial z^2} \tag{A.18}$$

In spherical coordinates

$$\overline{\nabla}\cdot\overline{V}=r^{-2}\frac{\partial(r^2V_r)}{\partial r}+(r\sin\vartheta)^{-1}\frac{\partial(V_\vartheta\sin\vartheta)}{\partial\vartheta}+(r\sin\vartheta)^{-1}\frac{\partial V_\varphi}{\partial\varphi} \tag{A.19}$$

$$\nabla^2U=r^{-2}\frac{\partial}{\partial r}\left(r^2\frac{\partial U}{\partial r}\right)+r^{-2}(\sin\vartheta)^{-1}\frac{\partial}{\partial\vartheta}\left(\sin\vartheta\frac{\partial U}{\partial\vartheta}\right)+(r\sin\vartheta)^{-2}\frac{\partial^2U}{\partial\varphi^2} \tag{A.20}$$

$$\left(\overline{\nabla}\times\overline{V}\right)_r=(r\sin\vartheta)^{-1}\left(\frac{\partial\left(V_\varphi\sin\vartheta\right)}{\partial\vartheta}-\frac{\partial V_\vartheta}{\partial\varphi}\right) \tag{A.21}$$

$$\left(\overline{\nabla}\times\overline{V}\right)_\vartheta=(r\sin\vartheta)^{-1}\left(\frac{\partial V_r}{\partial\varphi}-\frac{\partial\left(r\sin\vartheta\,V_\varphi\right)}{\partial r}\right) \tag{A.22}$$

$$\left(\overline{\nabla}\times\overline{V}\right)_\varphi=r^{-1}\left(\frac{\partial(rV_\vartheta)}{\partial r}-\frac{\partial V_r}{\partial\vartheta}\right) \tag{A.23}$$

$$\overline{\nabla}U=\left(\frac{\partial U}{\partial r},\frac{1}{r}\frac{\partial U}{\partial\vartheta},\frac{1}{r\sin\vartheta}\frac{\partial U}{\partial\varphi}\right) \tag{A.24}$$

Helmholtz's Theorem

A vector field $\bar{V}$ is defined in a simply-connected domain Ω, giving its divergence and curl in Ω as well as the normal component on the boundary Γ.

In a domain Ω bounded by Γ, given

$$\begin{aligned}
\overline{\nabla} \cdot \overline{V} &= \mathrm{s} && \mathit{in}\,\Omega \\
\overline{\nabla} \times \overline{V} &= \bar{c} && \\
\overline{V} \cdot \bar{n} &= \mathrm{h} && \mathit{along}\,\Gamma
\end{aligned}$$

the vector field $\bar{V}$ is defined in a unique way.

MIX
Papier aus verantwortungsvollen Quellen
Paper from responsible sources
FSC® C105338

If you have any concerns about our products,
you can contact us on
ProductSafety@springernature.com

In case Publisher is established outside the EU,
the EU authorized representative is:
Springer Nature Customer Service Center GmbH
Europaplatz 3, 69115 Heidelberg, Germany

Printed by Libri Plureos GmbH
in Hamburg, Germany